THE

SOYBEAN BOOK

Growing & Using Nature's Miracle Protein

Phyllis Hobson

GARDEN WAY PUBLISHING
CHARLOTTE, VERMONT 05445

Illustrations by Andrea Gray

Printed in the United States

First printing, April 1978

Library of Congress Cataloging in Publication Data

Hobson, Phyllis.
　　The soybean book.

　　Includes index.
　　1. Cookery (Soybeans)　2. Soybean.　I. Title.
TX803.S6H63　　　　641.6'5'655　　　　78-5594
ISBN 0-88266-130-2

Contents

Why Soybeans?

If you're looking for good nutrition, look no further. If you're searching for a high-protein food to take the place of meat in your diet, you've just found it. If you need information on a low-cost way to vary your family's meals, this is it.

Whatever your needs, your tastes or the condition of your wallet, the soybean—nature's miracle protein—is the answer. It can add taste, texture, and good nutrition to your menus at less cost than your daily cup of coffee. It can even supply an alternative to that beverage, too, if you like.

The only vegetable source of all the essential amino acids, the soybean is one of the most concentrated, nutritious foods known to man. Pound for pound, soybeans contain more protein, calcium and B vitamins than do meat and other animal protein foods. Two pounds of soy flour will supply the protein equivalent of five pounds of boneless beef, fifteen quarts of milk, six dozen eggs or four pounds of cheese.

Besides their value as a protein food, soybeans contain vitamins A, B, B_1, B_2, E and K, iron, phosphorous, and potassium.

Soybeans are a dieter's delight. Their low carbohydrate content makes them ideal for weight watchers, diabetics and others on starch-restricted diets. One serving (half cup) of soybeans has only about 100 calories, far less than a serving of meat.

Soybeans are low in cholesterol but high in polyunsaturates. They

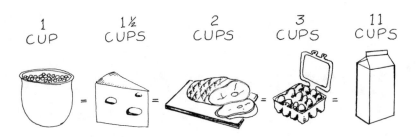

also contain a substantial amount of lecithin, a substance that helps prevent cholesterol build-up in the arteries.

Because there are losses—and gains—of vitamins in processing, different forms of soybeans have varying food values. Fresh soybeans, picked before the plant completely matures, abound in vitamin A, much of which is lost in the drying process. On the other hand, soybeans begin to develop vitamin C as they sprout, a process that continues after storage. Refrigerated soy sprouts are richer in vitamin C than tomatoes.

This little powerhouse of nutrition is easy to plant, grow, and harvest. An amateur gardener with a small backyard plot can grow enough soybeans to supply a family for a year. They require no special soil, fertilizers, or cultivating knowledge. You can process them with equipment already on hand, and store them indefinitely in a small space.

Soybeans rate tops in times of disaster. For a quick source of protein, a crop can be developed in sixty to seventy days, compared with six months for pork and one and a half to two years for beef.

A little bit of land goes a long way for soybean crops. An acre of land used to produce feed for livestock can provide one person's protein requirement for 250 days. That same acre can supply almost ten times that amount of soybeans—enough protein for one person for 2,200 days. And there is no other way to produce so much protein for so little energy. It takes one person 100 hours to grow enough livestock feed to produce forty-five pounds of beef. That same person could produce 1,000 pounds of soybeans in an equal period of time.

When it comes to versatility, soybeans really shine. You can serve them fresh, dried, whole, ground, powdered, and sprouted; as a meat, milk, cheese, vegetable, beverage, bread, or nut. Add them to meatloaves, candy, eggs, casseroles, salads, cereals, or flavor and serve them as a meat substitute. They're great for breakfast, lunch, dinner, or a snack. You could serve soybeans every day at every meal and go a long time without repeating a dish.

With their nearly flavorless quality, soybeans become chameleons in cooking. Roast them in the oven and they taste like peanuts. Beef flavoring can transform them into a hamburger loaf.

Their texture can suit your every whim. Put them through the blender with water and you have a liquid emulsion that looks and tastes like milk. Chop them, grind them, mill them, curd them. Soybeans can be baked, steamed, simmered, fried, broiled, or tucked into almost any food to add texture and good nutrition at low cost.

Soybeans are not a new or exotic food. One of the world's oldest crops, they were cultivated centuries before the building of the pyramids. For thousands of years they have been the one reliable source of protein for the working man of the Orient.

Explore the shelves of a supermarket today and you will find countless items containing soy products. Commercial food processors know the value of the inexpensive soybeans and have been adding them to your food since World War II.

Powdered soybean—called soy flour—has a great ability to absorb and hold moisture. Because of this attribute, it is used to give juiciness to bologna, wieners, and sausages. It gives substance and body to instant soup mixes, artificially flavored broths, and low cholesterol egg substitutes.

Soy flour is used as a dough conditioner to improve the flavor and browning quality of breads and rolls. It makes cookie dough easier to handle, increases the moisture content of cakes, reduces the amount of fat absorbed by doughnuts during frying, and keeps pancakes and waffles from sticking to the griddle.

Prepared mixes for cakes and pie crusts contain as much as 15 percent soy flour and millions of pounds of it are used each year in baby foods and in low calorie and dietetic foods.

Soy flour also is used in the crackers and snack foods you buy, in high-protein cereals, canned soups, noodles, macaroni and spaghetti, in non-dairy coffee creamers and whipped toppings. And soy nuts, in a variety of flavors, are available wherever snacks are sold.

Soy oils are sold for cooking and popping corn, in shortening, margarine, and salad dressings.

Ground soybeans are sold as a meat extender and are the main ingredients of the imitation meats to be found in health food stores.

In addition to uses as human food, soybeans are the main ingredients of dog and cat foods, of protein supplements for cattle, hogs, rabbits, and chickens, and of milk replacement for nursing calves and goats. The green soybean plant is used as a high-protein pasture grass for grazing livestock and for winter silage. Dried in the blossom stage, it makes an excellent hay.

Like everything else, soybeans aren't perfect. Because raw soybeans contain an anti-trypsin factor that inhibits the digestion of protein, they must be cooked before eating, no matter which form is used, to destroy that factor. This can be time-consuming.

Some people object to the firmness of soybeans. In the processing section we have included methods of soaking and cooking which soften soybeans somewhat, but on the whole, their firmness is something you cannot change. However, you can grind or mill them or convert them into soy curd or soy sprouts.

In this day of instant everything, soybeans need twenty-four hours of soaking and long, slow cooking or processing. There's no getting around it. If you're going to enjoy home-grown soybeans, you're going to spend some time in the kitchen.

But it's time and energy well spent. Compared to the months required to raise livestock for human food, or the hours the average worker must spend in the factory or the office in order to afford those instant foods at the supermarket, the few minutes spent giving a pot of soybeans an occasional stir or rinsing a crop of soy sprouts is pretty minor.

Ironically, the biggest disadvantage of soybeans is the price. It is so low that few take them seriously. It's hard to believe that anything that cheap can be worth much.

But they are. By learning how to grow and use soybeans you can:

- Add good nutrition to your diet.
- Reduce the risk of heart and arterial disease.
- Cut the cost of at least some of your protein by 90 percent.
- Follow starch-restricted diabetic diets without deprivng yourself of good food.
- Prevent protein deficiencies in vegetarian diets.
- Supply alternatives for wheat and milk in allergy diets.
- Provide an easily digested milk during treatment of diarrhea and other digestive disturbances of babies or the elderly.
- Provide a good source of the B vitamins for bland diets.
- Supply roughage needed for good colonic health.
- Furnish the family with body-building vitamins and minerals in any form they prefer.
- Offer dieters a wide variety of low-starch, low carbohydrate, low-calorie vegetables, main dishes, and desserts.

Get acquainted with the soybean and you will find a friend for your budget and your health. What's more, you can save trips to the supermarket. Right on your own small plot of land you can provide for your family's table and your livestock in the barn.

Read on. We'll tell you how.

How to Grow, Harvest, and Store Soybeans

The lowly soybeans deserve a place of honor in your garden. They're not only versatile, they're easy to grow. Although yields are highest in the midwestern, cornbelt states, soybeans can be grown in almost any climate in the United States, at almost any altitude, with a large or small amount of rainfall. As a quick rule of thumb, most varieties of soybeans can be grown easily in any climate or location in which corn grows satisfactorily. Other varieties have been developed for other soil and climate conditions.

Soybeans grow best in warm weather and usually are planted in the spring. Although the beans sometimes fail to develop evenly in very hot weather, soybeans may be grown even in the most southern parts of the United States during the cooler seasons. Once the plant is established, it can tolerate extreme weather conditions, drought, heavy rainfall, and even frost. Beans which are fairly well developed before a killing frost will continue to ripen.

Soybeans prefer a sandy loam well fertilized with lime, potash, and phosphoric acid, but the sturdy plant really isn't too fussy. Even a well-drained soil isn't necessary, although it doesn't like to stand in water.

Sow soybeans, whether for dry or fresh beans, hay or pasture, after all danger of frost is past in the spring. Plant the seed after the soil begins to warm, about the same time corn or tomatoes are planted in your area.

Soybean seed may be planted from early spring to midsummer, depending on the temperature. Soybeans to be used fresh may be planted as late as sixty days before the first fall frost. For pasturage, hay and soil improvement, plant soybeans until August 1 in the north, September 1 in the south.

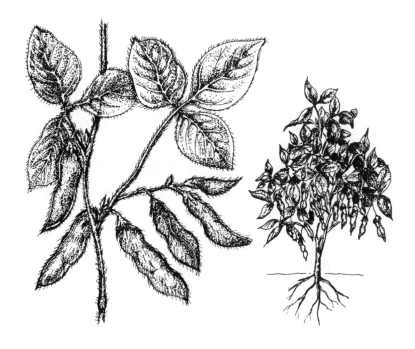

Selecting Seed

Much of the success of growing soybeans in the home garden depends on the variety of seed selected. In choosing a variety for your purpose, bear in mind that, in general, light-colored beans—yellow or green— have a milder taste. Darker varieties have a stronger, oilier flavor.

There are two distinct types of soybeans, field and garden varieties. The field beans are not suitable for table use and are available through farm seed dealers. Varieties of field beans have been adapted for the production of oil, for hay and pasturage, and for livestock grain supplements. However, the table varieties which taste better for humans also are suitable for animal feed.

There are hundreds of varieties of soybeans, developed for different uses, growing seasons, and soil and climate requirements. Some are old Oriental varieties which date back to the introduction of soybeans to this country from China in the 1800s. In the past twenty years varieties have been developed which make it possible to grow good crops of soybeans in the northernmost areas of the country.

All of the varieties can be divided into these three special-use groups:

Forage—Soybeans that produce an abundance of stems and leaves for hay and pasture feeding are comparable with alfalfa in food value. Generally these are the smaller-seeded black and brown late varieties, which have less oil than the commercial varieties.

Commercial—These varieties have high oil and protein content. Usually they are the yellow varieties.

Food—Edible varieties usually are larger, easier to cook and shell, and better flavored than other varieties. They may be any color.

For our purposes, we're interested only in the edible soybean varieties. Seeds of these varieties are sold in small, garden-size quantities by local dealers and through catalogs. Since soybean varieties often are developed for specific areas, look for those which are best suited to your climate and soil. Check with your local county agent or seed supplier before buying.

Here are some of the varieties available for the home garden:

Akita Early—A Japanese variety which produces creamy yellow dried beans in ninety-five days, fresh soybeans in sixty-five days.

Altona—An early soybean developed in Canada, this is a good choice for northern gardens. Seed is yellow with a black eye. It matures 100 days in the North, earlier in the South. It's ready for cooking as a fresh soybean in about seventy days.

Amsoy—One of the oldest American varieties. Dried beans mature in 100 days.

Envy—A high-yielding, bright green soybean developed in New Hampshire. It matures in northern gardens in 104 days; in about seventy days for fresh soybeans.

Fiskeby V—A patented plant developed in England for northern climates. A high-protein that provides fresh soybeans in sixty-eight days, mature beans in ninety-eight days.

Frostbeater—A heavy producer, Frostbeater matures in the most northern areas of the United States. Small plants produce fresh soybeans in seventy-five days, dried beans in 105 days.

Giant Green—One of the older, popular varieties, Giant Green produces a large size, green soybean which matures early for northern gardens.

Hakucho—A very early Japanese variety, Hakucho is adapted to short season areas.

Kanrich—A bush-type soybean, Kanrich is an old, very productive variety best adapted for midwestern climates, where it matures in ninety days and produces fresh soybeans in sixty days.

Meredith—The very tall plants develop small, dull-yellow beans. Developed in Maine, it matures in 110 days in that climate. Fresh soybeans may be harvested in eighty days.

Oriental Black—A Japanese variety, this black soybean matures in 100 days. Fresh beans can be picked in seventy days.

Panther—A black soybean popular in Japan because it is more easily digestible than many other soybean varieties. Mature beans take 115 days. Fresh beans are ready thirty days earlier.

Prize—A tall, erect plant adaptable to many parts of the United States. Each pod contains two, three, and occasionally four large beans. Fresh soybeans may be harvested in eighty-five days. Mature beans take 115 days.

Traverse—A high-yielding yellow soybean developed in Minnesota, Traverse matures in 111 days. Fresh soybeans are ready thirty days earlier.

Usually it's best to buy seed as close to home as possible to get the variety best suited to your climate. As a guide, early varieties should be grown in the North and medium to late varieties in the South. When in doubt about the best variety for your area, ask your seed dealer.

Sources of Seed

As soybeans gain in popularity with home gardeners, more and more large seed catalog operations are becoming interested in offering the seed in small quantities. For hay or pasture planting, you also may buy seed by the pound from farm seed suppliers in your area.

The following companies offer soybean seed in small quantities and will furnish seed catalogs on request:

Joseph Harris Co., Inc., Moreton Farm, Rochester, NY 14624.

W. Atlee Burpee Co., Warminster, PA 18974

Farmer Seed & Nursery Co., Faribault, MN 55021

J.W. Jung Seed Co., Randolph, WI 53956

Thompson and Morgan Inc., P.O. Box 24, Somerdale, NJ 08083

Nichols Garden Nursery, 1190 N. Pacific Highway, Albany, OR 97321

Henry Field Seed and Nursery, 407 Sycamore St., Shenandoah, IA 51602

Johnny's Selected Seeds, Albion, ME 04910

If you'd rather buy than grow your beans, soybeans for sprouting, milling and cooking as well as soybean flour can be ordered from:

Walnut Acres, Penns Creek, PA 17862

The Natural Development Co., P.O. Box 215, Bainbridge, PA 17502

Farmer Seed & Nursery Co., Faribault, MN 55021

Applewood Seed Co., 833 Parfet St., Lakewood, CO 80215

How Much Seed

Whether you are growing soybeans for table use or as a protein supplement for livestock, the method of cultivation and the variety planted may be the same. The amount of seed and the size of the soybean patch will differ, however.

For the small garden, a packet of seed (usually 100 to 150 seeds) will plant a twenty-five-foot row. For the larger garden, allow one pound of seed for each 150-foot row. If you're introducing soybeans into your family's diet, one 150-foot row should be sufficient the first year. As family members develop a taste for soybeans and as the cook develops skills in preparing them, perhaps you will want to increase the crop to one row per family member. Under good conditions you can expect a yield of fifteen pounds of soybeans from each 100 feet of soybean plants.

Planting the Seed

Prepare the seedbed for soybeans as you would for corn. It is not necessary to add commercial fertilizer before planting soybeans in rich soil or soil to which manure, barn bedding, or a variety of organic materials regularly are added. Soybeans need a soil with a pH of from six to seven. Where organic fertilizers are not used, it is advisable to add some form of limestone plus one-half pound of a commercial fertilizer such as 5-10-10 for every ten square feet of soil.

Fertilizer also may be side-dressed later, as the plants develop, by spreading a thin layer of commercial fertilizer beside the row of plants.

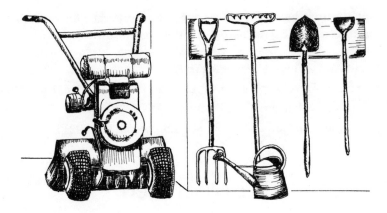

Keep any side dressing two to three inches from the plants to prevent damage.

Plant seed one to one and one-half inches deep, two to three inches apart in rows two to three feet apart. The distances will vary according to the variety and the method of weeding to be used. Smaller, earlier varieties may be planted in rows two feet apart if you plan to hoe by hand. Plant larger, medium to late varieties—or smaller varieties which are to be tilled by machine—in rows thirty to thirty-six inches apart.

Inoculation

Like other legumes, soybeans take care of their own nitrogen needs. Bacteria which live in the nodules on the plant's roots convert nitrogen from the air to a form the plant can use. There is always a surplus left in the soil which improves the soil for next year's crop.

However, the soybean only takes care of the bacteria's housing needs. It doesn't produce bacteria. The bacteria must be present in the soil when the soil is planted or they will have to be supplied by a process called *inoculation*.

If soybeans have been grown successfully in the soil for the past two years, you can be sure the bacteria are present in the soil and nothing will have to be done before planting the seed. But if soybeans have not been grown in the soil or if those grown there previously had greenish-yellow leaves, you'll need to inoculate the seed before it is planted. After you have grown two good crops of soybeans in that location, the seed need not be treated for subsequent crops.

Seed may be inoculated with soil taken from another part of the garden or a field in which soybeans (any variety) have been grown successfully. Or you may use commercial inoculant powder, which is sold by seed dealers. Be sure to buy inoculant made specifically for soybeans. Soybeans cannot be inoculated with bacteria from any other legume.

To coat the soybeans with the bacteria-rich soil or commercial powder, first moisten the seed with water in which sugar or molasses has been dissolved. One-half cup of sugar dissolved in one quart of warm water will moisten up to one bushel of seed. Spread the seed out and dust it with the amount of commercial inoculant listed on the package or with one gallon of finely sifted soil. Mix well by hand or with a hoe until the seeds are well coated.

Soybeans that are supplied with an abundance of the bacteria needed to use the nitrogen in the air produce larger plants and more beans of a higher protein value. It makes no difference whether the bacteria was already in the soil or was introduced by the gardener.

Germination

After the seed is planted, it takes from five to ten days for most soybean varieties to germinate, according to the moisture content and the temperature of the soil. From the time they germinate, if the soil is warm, the little plants will grow quickly, thrusting the cotyledons up through the soil, then adding leaves. Very soon you'll see distinct rows of soybean plants.

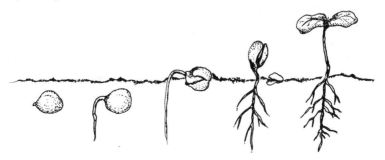

Within two weeks after germination, the plants will be four to six inches tall and should be thinned to three or four inches apart to make room for the growing plants. Keep the rows free of weeds to give the plants room to grow and to allow them the nourishment they need without competition.

At this point you may have a little competition yourself. Soybeans seldom are bothered by insect pests or disease, but rabbits, woodchucks, and other small animals relish them and may harvest your crop before it develops.

There are several ways to keep rabbits out of the soybean patch. If you have blood or bloodmeal available (possibly from a local butchering plant) just spread a thin ring of it around the soybeans. The smell frightens them away and the blood nourishes the soil. If there's a zoo nearby, try to get a bucket or two of lion manure. It's said that this frightens away small animals. Lacking either of these, you can use a simple physical barrier. Fence in your soybean patch, to keep out small animals.

The Developing Plant

As the young soybean plant grows, it develops trifoliolate leaves (leaves in groups of three). By the time the plant is twelve to fourteen inches tall, it has five or six of these groupings and is ready to bloom. Flowering times will vary because the soybean plant is unique: the time of blooming depends on the length of the days rather than on the temperature. Once it begins, blossoming will continue for about three weeks while the plant grows to its full height—twenty-four to thirty-six inches, according to the variety.

Now the pods begin to form, small at first and covered with the typical soybean fuzz. As they grow, the outline of the three beans inside begins to show.

Under the ground, the nitrogen-fixing nodules are forming on the roots of the plant. Each nodule contains millions of bacteria which convert nitrogen from the air into a form usable by the plant. These nodules not only feed the plant, they spill over and enrich the soil.

The bean pods fill rapidly and as the plant approaches maturity, the leaves on the bottom of the plants turn yellow. The pods should be

plump and green. At this point, they are ready to be picked for fresh soybeans. Pick only those beans which are the plumpest. The others will develop, extending the fresh soybean harvest for about two weeks.

After the soybean has passed the fresh bean stage, the leaves will turn yellow and begin to drop off, from the bottom up. Then the stalk and bean pods will shrivel and turn brown. When the plant is brown and bare of leaves, with only the dried pods left on the stalk, it is ready for harvest for dried soybeans. If the beans are allowed to stand too long after they reach this stage, the pods may shatter and the beans will be lost. There is danger, too, of the pods molding in rainy weather.

The Harvest

If you've grown only a small amount of soybeans, they may be harvested just as you would any dried beans, by pulling the plant and plucking off the pods, which may then be shelled at leisure in a comfortable spot in the shade.

Larger amounts may be cut with a knife and stacked in small shocks for curing and drying. Stand the shocks upright so the pods do not touch the ground and leave them four or five days in the sun, or in a shed if the weather is wet. Turn occasionally for uniform drying.

Removing the dried soybeans from the pods isn't as simple a matter as shelling other dried beans. Dried soybean pods have a bulldog tenacity. It is possible to shell small batches by hand, and for commercial field harvests, the combine is a necessity, but to thresh a few bushels, you can use one of these methods:

1. Pull or cut the dried stalks by hand and cure a few days under cover to make sure they're dry. Thresh five or six stalks at a time by holding the end of the stalks and beating them back and forth against the inside of a deep box or barrel. Most of the beans and some of the chaff will fall to the bottom of the container and the stalks with the empty pods may be returned to the garden or placed on the compost pile after you've gone over them by hand to remove any pods not shelled.

2. Pluck the pods from the stalks and spread the dried pods out thinly on a clean tarp or canvas. Flail them with a stick, carpet beater or tree limb, beating them well to break open the pods and release the beans.

3. Pluck the dried pods from the stalks and spread them out on a tarp or canvas, then stomp the beans out by walking back and forth over them with your shoes on. A good bet for this method is to talk some youngsters into dancing on the beans. Soybeans are pretty hard, so there isn't much danger of crushing them.

When most of the beans have been removed from the pods, much of the chaff and dried pods may be scooped off the top and discarded. Any remaining chaff may be removed by pouring the beans from a height of five or six feet in front of a fan. The heavier beans will fall straight down to a container on the floor while the fan will blow the chaff and debris to the side. It takes several passes in front of the fan to remove most of the chaff.

An easier but riskier way of removing the fine chaff is to wash the beans in a large container, such as a wash tub or boiler. Fill the container no more than one-third full of beans and chaff, then pour in water to overflowing. Keep the water pouring as the container overflows. Stir occasionally. The beans will stay on the bottom, but the chaff will float to the top and be washed away as the water overflows. This method should not be used unless you are sure you can dry the beans quickly. Otherwise they will sprout and mold.

Storing Beans

Before storing the soybeans, you should be sure they are sufficiently dry. The ideal moisture content for storage is 13 percent. If you live near a grain elevator in a farm area, see whether the elevator offers the service of measuring moisture content for a small charge. If not, you can test for moisture by placing a soybean on a firm surface and tapping it sharply but lightly with a hammer. A well-dried bean will split neatly in two.

If the soybeans are not dry enough for storage, you may finish the drying process by spreading them out on drying trays in the sun or on tarps or sheets of black plastic on a flat surface. A driveway does nicely. The trays may be carried, and the tarp or plastic may be picked up by the four corners to form a bag in which to move the beans inside at night. Larger amounts may be spread thinly on a clean, wooden floor in a shed for a week or two and stirred once a day with a rake. Test again before storing.

Thoroughly dry beans may be stored in loosely covered plastic containers or in burlap bags kept in a well-ventilated place. Soybeans should not be covered tightly.

Stored soybeans seldom are attacked by weevils, but they should be kept out of reach of mice and rabbits. A new metal or plastic garbage can with a loose-fitting lid is a good mouse-proof container and will hold a year's supply of soybeans for the average family. If they are kept dry and clean, dried soybeans will keep well for several years.

Saving Seed

While you're growing your own soybeans in the garden, it's a simple matter to save enough seed for next year's crop.

It's best to start at the beginning, when you plant soybean seed in the spring. Set aside an area as your seed-producing garden. You'll need nearly ten feet of row for every pound of seed you want to save. By saving the first beans from the plants grown in this area, you will be developing your own early variety. You may then use the later beans for food or save them also for seed. First though, you will be assured a supply of seed for next year.

If you are blessed with a cool, dry autumn, all you need do is leave the beans on the vines until they are completely dry, then pick the dried pods. It's best to shell them by hand to avoid any chance of damaging the beans.

If the season is short or the weather is wet, you can pull the stalks when the pods have all turned yellow. Stack them loosely, standing upright, inside a shed or barn until the pods are brown and shriveled and the beans inside are dry.

The pods also may be spread out on trays or on a tarp or sheet of plastic in a cool, dry place—not in the sun or artificial heat. The dried beans may be removed from the pods any time before spring.

Don't worry about removing the fine chaff. Next spring it can be planted in the ground right along with the seed.

Before storing, spread the shelled beans thinly in trays or on any flat surface in a cool, well-ventilated place for a few days, stirring once or twice a day.

Store the dried seed in cloth bags, cardboard boxes or glass jars with cheesecloth fitted over the opening as a lid. Soybean seed should not be stored in airtight containers, should not be subjected to temperatures above 70° F. or below 32° F. and should be kept dry.

Well-dried soybean seed should maintain its viability for two or three years, although germination is highest (80 to 85 percent) the first year. During the second year, germination may drop to 65 percent but the seed may still be worth using.

Some seeds will die during storage, even under the best conditions. Before planting, the seeds should be tested for germination. To do so, take a small handful of seed and discard any broken, split or obviously dead seed and any that are small or abnormally formed. Put the rest in a bowl or enamel pan and cover with lukewarm water. Discard any seeds that float to the top of the water.

Saturate a clean dishcloth or dish towel in warm water and squeeze out the excess water. Spread out the cloth, count out twenty-five seeds and spread the seeds over the damp cloth. Roll it up, seeds inside, and keep in a warm (70° to 80° F.) place. Sprinkle with water as needed to keep cloth damp.

After five or six days, unroll the cloth and count the seeds which have germinated. If twenty to twenty-two seeds have sprouted, germination is excellent—80 to 88 percent. If fewer than fifteen beans have sprouted, germination is less than 60 percent and the seed should be discarded. In between these two figures, whether or not to use the seed is up to you. You might consider using it but planting the seeds 1½ to 2 inches apart in the rows.

Putting the Garden to Bed

Once your soybean crop is safely stored for the winter, you should finish the growing season with one final step which will prepare the garden for next year's crop.

First supply the soil with lime by sprinkling it with wood ashes, crushed egg shells (save them all year for this use), or agricultural lime, then till or shovel under the empty soybean pods and dried stalks still standing in the garden. Over the winter they will be absorbed by the soil and supply it with the needed nitrogen, phosphoric acid, and potash at no cost.

About Next Year's Garden

Next year, plant your soybeans in another location in the garden, if possible. By rotating this soil-building crop, you will benefit the entire garden.

How to Process Soybeans

Fresh Soybeans

Fresh soybeans are the immature vegetable, picked as soon as the bean inside the pod has reached full size, but before it has toughened and begun to dry.

Fresh soybeans may be cooked and used as a green vegetable, much like green lima beans. They not only have a rich, nut-like flavor and are delicious alone or combined with other vegetables, but their high protein content is a good meat substitute which makes them a welcome addition to the diet.

Fresh soybeans contain about 600 calories to the pound and only about one half the carbohydrates of other beans. They are a good source of vitamin A, the B vitamins and some vitamin C.

Cooked fresh soybeans should be used as a hot vegetable, in salads or in vegetable dishes for meals which might otherwise be lacking in protein—in meals, for instance, in which cheese or cottage cheese are the main protein source. Fresh soybeans also are an excellent, high-protein addition to soups and meat-stretching casseroles.

Pick them when the pods are plump and green, before the leaves turn yellow. Most varieties remain in this stage less than two weeks. so you'll need to can or freeze any surplus to prolong the season.

Handle fresh soybeans quickly after picking. They begin to lose flavor almost immediately and should be shelled and cooked within two hours of picking.

Even at this stage, soybean pods are tough and hairy, making the beans hard to shell. To simplify the process, cover the green pods with boiling water and soak them for five minutes. Drain. Remove the beans by holding the pods between the thumb and forefinger and snapping the pod in half crosswise. Squeeze out the beans. Cook shelled beans in a small amount of salted water ten to twenty minutes, or until tender. Serve immediately. One pound of soybeans in the pods will yield one-half pound of shelled beans, or about one cup.

The soybeans may be cooked right in the pod. Just drop the clean pods into boiling water, cover and cook twenty-five to thirty minutes. Cool, shell, and reheat before serving. They also may be cooked and served right in the pod and eaten like artichokes. Each diner shells his own by holding the pod with his fingers while biting down gently on the pod with his teeth. A tug on the pod with the fingers and the shelled beans end up in the mouth.

To Freeze Fresh Soybeans

Process no more than two pounds of pods at one time. Cover freshly picked soybeans with boiling water and soak for five minutes. Shell. Cook five minutes in a small amount of water. Cool and package, then freeze in pint or half-pint containers.

To Can Fresh Soybeans

Allow four to five pounds of soybeans in the pods for each quart jar. Shell as above. Cover the shelled beans with hot water and bring to a boil. Boil three minutes. Pack hot into hot quart or pint jars. Cover with cooking water, leaving one inch of head space. Adjust lids on jars

and process pints fifty-five minutes, quarts sixty-five minutes at ten pounds pressure. Because they are a low-acid vegetable, soybeans must be processed in a pressure canner.

Dried Soybeans

Cooked, dried soybeans are a protein-rich, heavy vegetable and are best used sparingly, in combination with grains or other vegetables.

They may be mashed to a pulp and added to meat loaves and meat patties, mixed with eggs and vegetables and sandwich spreads, blended into bread and cookie batters, or cooked as a meat substitute.

Dried, uncooked soybeans are roasted and eaten as a snack, milled into flour and used in almost any batter, diluted in water and used as a beverage or milk substitute, sprouted, and eaten as a fresh vegetable rich in vitamin C.

Remember, soybeans must be cooked—steamed, boiled, simmered or roasted—whether they are to be used fresh or dried and whatever the dried form may be. Cooking is necessary to destroy the anti-trypsin factor which makes soybeans hard to digest.

SOAKING

Dried soybeans are firm, compact beans which do not soften as much as other dried beans, even after long cooking. To improve their flavor and texture, they must be soaked before cooking. Here are the two best methods of soaking:

1. Cover soybeans with cold water and soak twenty-four hours in the refrigerator. Refrigeration is necessary to keep the beans from fermenting, for soybeans ferment much easier than do other beans.

2. Place soybeans in a flat, shallow pan such as a cake pan or ice cube tray. Cover with cold water, refrigerate and soak eight to twelve hours, then freeze overnight or until water is solid. It is not necessary to thaw them before cooking.

COOKING

After soaking, dried soybeans may be cooked in a saucepan, in a pressure cooker, or in an electric slow cooker. Do not drain off the soaking water before cooking. It contains many of the nutrients from the beans. Do not add soda to cooking water. Soda destroys B vitamins, a valuable nutrient of soybeans. More water may be added, if needed, but the soaking water should be used for cooking.

SAUCEPAN METHOD

To cook dried soybeans in a saucepan, heat soybeans and soaking water to the simmering—not boiling—point. Lower heat and cook slowly until beans are tender, four to five hours, adding water as needed. Be sure to keep the heat low. Soybeans are a high-protein food and high heat toughens their protein just as it does the protein in steak. The secret to tender soybeans is long soaking and long, slow cooking.

PRESSURE COOKER METHOD

To cook dried soybeans in a pressure cooker, add only enough water to barely cover the beans after soaking. Do not fill the cooker more than one-half full. Following manufacturer's directions, cook at fifteen pounds pressure for one hour. Let the cooker cool by itself and be sure all pressure has been released before opening.

SLOW COOKER METHOD

Soak dried soybeans in just enough water to cover. If cooker is adjustable, bring just to simmering at high heat, then lower and cook eight to twelve hours on low heat. If it is not adjustable, or if you prefer, cook on low heat twelve to twenty-four hours.

SEASONING SOYBEANS

Some people like soybeans just as they are the first time they eat them. Others acquire a taste for them over the years. But a lot of people believe that soybeans alone are flat and tasteless. If you are among the latter group, or if you're still acquiring your taste for soybeans, you might try adding a little more salt than usual or pepping them up a bit with the addition of flavorful vegetables and seasonings. Most of the seasonings available are suitable for vegetarians.

The following recipes are examples:

FLAVORED DRIED SOYBEANS

1 cup dried soybeans
3 cups cold water
½ teaspoon salt

2 tablespoons vegetable broth
    powder or granules
3 tablespoons celery, chopped
1 tablespoon soy sauce

Soak soybeans in cold water according to earlier directions. Add remaining ingredients and cook in pressure pan forty-five minutes at fifteen pounds pressure. Let pressure fall of its own accord. Serves 6.

SOYBEANS IN TOMATO SAUCE

1½ cups dried soybeans   1 cup chicken stock or vegetable
2 cups cold water   cooking water

Soak soybeans in cold water according to earlier directions. Add
chicken stock or vegetable cooking water and cover pan. Simmer four
hours, adding more liquid if needed. When tender, add:

1 tablespoon salt   1 cup tomato puree
1/8 teaspoon pepper   1 tablespoon Worcestershire
2 tablespoons vegetable oil   sauce
2 cloves garlic, minced   1 bay leaf
1 onion, chopped   2 tablespoons parsley, chopped

Simmer, uncovered, 30 minutes longer. Serves 6 to 8.

Freezing and Canning

Dried soybeans store well without processing. However, for conven-
ience and ease in cooking, they may be frozen or canned. Here are two
ways:

QUICK-COOKING DRIED SOYBEANS

Soak dried soybeans in cold water according to earlier directions. Heat
to just under the boiling point, turn off heat and wait until cool. Package and freeze. To use, cook (don't thaw in advance) for one or two
hours, until tender.

SOYBEANS CANNED IN
TOMATO SAUCE
(Taste much like canned pork and beans)

2 pounds dried soybeans	2 teaspoons salt
2 quarts cold water	1 cup onions, chopped
1 quart tomato juice	¼ teaspoon powdered cloves
3 tablespoons sugar	¼ teaspoon powdered allspice

Soak soybeans in cold water according to earlier directions. Simmer beans for two hours in soaking water. Drain, reserving liquid, which may be used to add protein to soups and stews. Fill canning jars about three-fourths full of drained soybeans. Meanwhile, in a saucepan, combine tomato juice, sugar, salt, onions, and spices. Bring to boil. Pour hot sauce over beans, filling jars to within one inch of top. Adjust lids and process pints 65 minutes, quarts 75 minutes at 10 pounds pressure.

Yield: 6 pints.

Soy Grits

To speed cooking and to produce a meat-like texture in some dishes, soybeans may be ground coarsely before cooking. These ground, then cooked soybeans are called soy grits.

Grits may be ground in any of several ways—with a hand crank or electric food grinder, using the coarse or medium blade, by a quick run-through a food blender, or with a soybean mill adjusted for cracking grains.

Coarse soy grits are soybeans which have been cut into eight to ten pieces. Medium grinding may vary according to your preference and the use you plan to make of the grits.

Cooked soy grits may be used in a variety of casseroles and meat or meat-substitute dishes. Because they have a meat-like texture and little flavor of their own, soy grits may be seasoned and used with or without meat in main dishes. To add a meat-like flavor, add soy sauce, Worcestershire sauce, or any of the broth powders or granules available in supermarkets and health food stores.

Because of the versatility of grits, many of the dishes in the recipe section call for cooked soy grits as an ingredient.

COOKED SOY GRITS

1 cup dried soybeans   Water

Soak dried soybeans in water to cover according to earlier directions. Drain, reserving soaking liquid. Coarsely grind beans through medium blade of food chopper or in blender. The ground beans should resemble cream-style corn. They may be cooked in a covered saucepan, in a pressure cooker, or electric slow cooker.

Saucepan: Add four cups water (including the reserved liquid) and one teaspoon salt to soy grits. Cook two to three hours, stirring occasionally, until soy grits are tender and almost dry.

Pressure Cooker: Add one cup soaking liquid and one teaspoon salt. Cook fifteen minutes at fifteen pounds pressure, following manufacturer's directions.

Slow Cooker: Add back one cup soaking liquid and one teaspoon salt. Heat just under boiling on top of stove, then pour into slow cooker. Cover and cook over low heat four to six hours.

Yield: about 3 cups cooked soy grits.

SEASONED SOY GRITS

3 tablespoons butter or margarine
2 tablespoons onion, minced
3 cups cooked soy grits
   (previous recipe)

1 tablespoon beef- or chicken-
   flavored broth powder or
   granules
1 cup tomato juice
1 tablespoon soy sauce

Melt butter and add onion. Cook over low heat, stirring frequently, until onion is transparent. Add remaining ingredients and mix well. Cook over low heat until almost dry. Serve as a meat dish. Serves 6.

Soy Pulp

Some of the dishes in the recipe section call for soy pulp, which is the cooked, mashed soybean with the outer covering of the bean removed.

To make soy pulp, soak and cook soybeans according to the directions earlier in this chapter. Cool slightly, then force through a food mill or colander to mash the pulp and remove the hulls.

For every two-and-a-half cups of soybean pulp desired, you'll need about one-half cup of dried, uncooked soybeans.

Soy Flour

Soy flour is not a true flour, but is a protein-rich concentrate, more like powdered milk or powdered egg than wheat flour. However, it may be used as a flour in bread, cake and cookie doughs.

Soy flour is unlike wheat flour in many ways. The fine powder made by milling soybeans has practically no starch, nor does it have the thickening properties of wheat flour. It has almost no gluten, so it cannot be used alone to make yeast doughs. For best results, use no more than one-fourth cup soy flour per cup of wheat flour. A small amount will keep wheat bread soft and moist, but too much will make the dough heavy.

However, in some doughs which do not depend on gluten for lightness, such as cookies, some cakes and pancake batters, as much as half of the wheat flour may be replaced with soy flour. It is best to begin cautiously and gradually increase the amount of soy flour.

Soy flour is difficult to measure consistently because it compacts. The most accurate way to measure is by weighing it on a kitchen scale, but it is possible to obtain good results by being careful to measure soy flour in the same way each time.

In addition to its uses as a flour, finely powdered soybeans may be used to make a form of soy milk, as described later.

How to Make Soy Flour

Soak the dried soybeans overnight in cold water to cover. (Refrigerate as usual.) In the morning, stir well and skim off any hulls that float to the surface. Simmer ten minutes. Drain through two thicknesses of cheesecloth, reserving the liquid for cooking uses.

As soon as the beans are cool, spread out on cloth-covered metal trays, on wire screen racks, or in a food drier. At this point, quick drying is important to keep the beans from sprouting or spoiling. Place the trays or racks in direct sunlight where circulation of air is good. Dry beans completely, taking them inside at night to protect them from dew. Finish drying by lightly toasting them fifteen to twenty minutes in a 200° F. oven. Watch carefully to keep beans from browning.

To test for proper drying, hit a few beans sharply but lightly with a hammer. The beans should break cleanly in two, with the hull separating from the meat.

The next step is to crack the dried beans by grinding them through the coarse blade of a food chopper. This loosens the hulls for winnowing.

To remove the hulls, slowly pour the cracked beans from a six-foot height in front of a fan. The beans should drop into a tray or basket at floor level as the hulls are blown to the side. Two or three passes may be necessary to remove all the hulls.

You now have a coarse bean meal which is ready to mill into flour. Using a hand grinder or a flour mill designed for soybeans and following the manufacturer's directions, grind the beans to a fine powder. Four pounds of raw beans will make more than three pounds of soy flour.

Soy flour, like whole grain flours, is at its peak of nutrition and flavor when used as soon as possible after milling, so it is best to mill it in small batches. However, small amounts may be kept tightly covered in the refrigerator or freezer for use in daily cooking.

How to Use Soy Flour

A small amount of soy flour will add nutrition to almost any recipe that calls for wheat flour. A safe rule is to substitute two tablespoons of soy flour for an equal amount of wheat flour for every cup of wheat flour used.

Soy flour browns more easily than wheat flour, so it's a good idea to lower the oven temperature by twenty-five degrees.

In adjusting recipes to soy flour, add a little more salt and a little more liquid to the ingredients.

In all the recipes in this book, the term "wheat flour" refers to whole wheat or white, all-purpose wheat flour, whichever is preferred. For recipes in which it is important to use either whole wheat or white flour, the kind of wheat flour is specified in the list of ingredients.

Sources of Equipment

Equipment for grinding soybeans into flour ranges from inexpensive hand grinders to the more expensive electric mills manufactured specifically for that purpose.

Because of their high oil content, soybeans will clog and ruin the stones of most stone-burr grain mills. Check to be sure the grinding equipment you choose is suitable for grinding soybeans.

The following companies sell grinding equipment made especially for making soy flour in the home. They will send descriptive material on request.

GARDEN WAY CATALOG, 1300 Ethan Allen Ave., Winooski, VT, 05404. Offers Marathon Uni-Mill, an electric, all-seed mill which adjusts from soybean splitting to a fine flour. Also offers a hand-cranked steel-burr grain mill suitable for soybeans.

NELSON AND SONS, INC., P.O. Box 1296, Salt Lake City, UT, 84110. Offers a hand-cranked, adjustable Quaker City mill which is better for coarse grinding than for fine flour milling.

THE GROVER Co., 2111 S. Industrial Park Ave., Tempe, AZ, 85282. Manufactures Marathon Uni-Mill, an electric mill which will grind any kind of grain, legume, or seed.

SUNSET MARKETING, 10330 Fair Oaks Blvd., Fair Oaks, CA, 95628. Offers three grain mills recommended for soybeans: Excalibur (available in kit or completed form), Marathon Uni-Mill and Grind-All, a steel burr mill.

ALL-GRAIN DISTRIBUTING CO., 3333 S. 900 East, Salt Lake City, UT, 84106. Offers All-Grain electric adjustable flour mill.

NICHOLS GARDEN NURSERY, 1190 N. Pacific Hwy., Albany, OR, 97321. Offers Corona Wheat Mill, a hand-cranked model adjustable for coarse or fine grinding.

Soy Milk

Soy milk, a creamy-white emulsion made from dried soybeans, resembles milk in appearance, taste, and consistency. It can be used in any way cow's milk is used, as a beverage and in cooking.

However, soy milk has less calcium than cow's milk, so if it is used exclusively, another source of calcium should be sought. Soy milk contains as much protein as cow's milk, in addition to iron, phosphorous, vitamin A, and the B vitamins.

Soy milk is an excellent substitute drink for those who are allergic to cow's milk or to whom cow's milk causes digestive upsets. It also is a good alternative for vegetarians who choose not to eat animal products of any kind.

It is an inexpensive, conveniently stored form of milk for emergency use, since the dried soybeans may be stored indefinitely, ready to make soy milk as needed.

When prepared at home, a quart of soy milk costs less than one-fourth the price of a quart of cow's milk purchased at the grocery store. Even if the milk comes from the family cow, it is far cheaper to raise a few soybeans than to feed a cow.

Once made, soy milk should be treated just as you would fresh milk. It should be cooled as quickly as possible and kept refrigerated. It is as perishable as any milk without preservatives and will sour if not used in a few days. It should be kept tightly covered, for it will absorb odors and flavors as will cow's or goat's milk. When heated, it must be watched carefully to keep it from scorching.

Soy milk may be substituted for cow's milk. It adds a creamy, satisfying consistency to white sauce and creamed soups and may be used to make a cheeselike food called soy curd. It can be curdled with vinegar for use in recipes calling for sour milk. A small amount of soy flour may be added to soy milk to give it a richer, heavier consistency.

There are many recipes for making soy milk, but these are the two we found most satisfactory. The first recipe is a rather involved process, but the product has a flavor and texture more like cow's milk than any other tried. The second recipe is quicker and simpler and is entirely satisfactory for cooking.

Both recipes call for sugar as a flavoring. (Cow's milk naturally has a like amount of milk sugar.) An equal amount of honey or molasses may be substituted if the milk is to be used within a day or two and is to be used for drinking only, but soy milk made with honey or molasses will not keep well and should not be used for making yogurt or soy curd.

Each recipe makes two quarts of soy milk. They may be adjusted to make enough at one time for three or four days. If refrigerated, soy milk will keep that long and it takes no more time to make one gallon than it does to make one quart.

SOY MILK
(Makes two quarts)

½ pound dry soybeans   ¼ cup sugar
1 quart cold water   ½ teaspoon salt
⅓ cup cooking or salad oil

Soak soybeans in cold water according to earlier directions. Processing one-half cup at a time, put soaked beans and soaking water through medium-coarse blade of blender or food chopper, pouring the resulting puree into a large enamel pan. Heat slowly to just below boiling. Remove from heat and let set thirty minutes. Strain through three thicknesses of cheesecloth, reserving both liquid and pulp. (Pulp may be added to meat loaves or casseroles.) Again bring the strained liquid to just below the boiling point and simmer 45 minutes over low heat. Keep heat low and stir often to keep from sticking. Remove from heat, add oil, sugar, and salt and run through blender at high speed. Return to heat and simmer fifteen minutes more. Cool quickly and add water to make two quarts. Keep refrigerated.

SOY MILK
(From soy flour)

8 cups cold water
2 cups soy flour
  (See earlier recipe.)
4 tablespoons sugar

4 tablespoons cooking or salad
  oil
½ teaspoon salt

Gradually stir water into soy flour. Mix well and let set two hours. Heat in the top of a double boiler to the simmering point, then lower heat, cover, and cook 40 minutes. Cool slightly and strain through three thicknesses of cheesecloth. Add sugar, oil, and salt. Mix thoroughly in blender. Add cold water to make two quarts. Chill before using.

Soy Curd

Soy curd is an unripened cheese made of soy milk. It is also known as soy cheese, or by its Japanese name, *tofu*. In the Orient, where for centuries the soybean has been known as "the poor man's cow," soy curd has been called "the meat without bones."

Soy curd is made from soy milk in much the same way soft cheese is made from animal milk. The soy milk is curdled by adding mineral salts or an acid. After the curds have formed, the liquid is strained through cheesecloth and the curd is used as a soft cheese or pressed into a firm cake.

Fresh, unpressed soy curd is rather soft and watery. It may be mashed and seasoned and used like cottage cheese in salads and main dishes. Fresh curd also may be flavored and used much as you would cream cheese in salads, sandwiches, and desserts. It is very perishable and must be covered and refrigerated.

For a firmer bodied cake, soy curd may be steamed or boiled twenty to thirty minutes in a cheesecloth bag. Cooked soy curd may be used as a starch-free noodle or macroni ingredient in casseroles or soups.

The fresh curd also may be pressed to remove more of the water and to give the cake more body. Pressed curd may be soaked in salt water, then cut into small pieces and dried to a semi-hard consistency. Dried curd has a sausage-like appearance and may be sliced, then used in cooked dishes. Pressed or cooked soy curd, sliced thin or diced, is a nutritious addition to almost any soup.

Fresh soy curd is very perishable and should be freshly made the day it is to be served. However, left over curd may be covered or immersed in water and kept in the refrigerator three or four days. Curd that has been cooked or pressed will keep a week or more in the refrigerator or it may be preserved by freezing, salting, smoking, or deep-fat frying. Dried curd will keep several weeks.

Making Soy Curd

To make soy curd, curdle soy milk by leaving it in a warm place until it sours and thickens. If you prefer, add lactic acid, vinegar or lemon juice to hasten coagulation. Add one tablespoon acid for each quart of soy milk.

When the milk is thick, cut it into chunks with a knife and set the container in a pan of lukewarm water. Slowly heat the water to the boiling point. The heat should be so low that it takes thirty to forty minutes to reach this point. Turn off heat and let set ten minutes, then strain curds through a cheesecloth-lined strainer. Tie cheesecloth into a bag and let hang until curds are as dry as possible.

You now have fresh soy curd which may be seasoned with salt, pepper, and maybe a little sour cream.

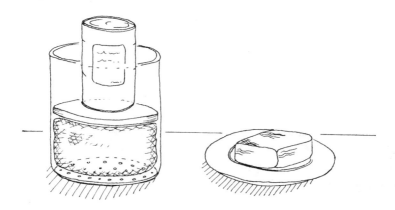

For a firmer curd, leave the fresh curd tied in the cheesecloth bag and steam or boil it in water to cover twenty to thirty minutes. Cool, remove the cheesecloth, and slice or serve according to the recipes in the recipe section. Small, soft cakes of curd may be dropped in the water and cooked until they float to the top.

You may press soy curd as you would cheese made from cow's milk. An inexpensive cheese press may be made by punching holes in the bottom of a metal coffee can with a nail. The fresh curd, still in its cheesecloth bag, is inserted into the can and followed by a saucer, small plate or wooden disk that fits the open end of the can. A weight (filled jars or cans of food work well) is placed on the plate to squeeze the water from the curd. The water drains out through the holes. Let set three to four hours or until curd is firm enough to be handled. For a firmer curd, use more pressing weight.

Soy curd also may be pressed on the kitchen drainboard. Place a sheet of waxed paper on the drainboard, then place one or more soy curd cakes on it. Cover with a second sheet of waxed paper and a wooden cutting board. Weight the top with heavy objects, such as books or cans of food. Let set five hours or overnight.

To dry soy curd, cut the cooked or pressed curd into thin slices and spread the slices out on trays which have been covered with cheesecloth or wrapping paper. Partially dry in a sunny, well-ventilated spot. When dried, the curd should have a crusty exterior, but still be soft inside. It must be kept in the refrigerator, but keeps longer than cooked or pressed curd.

Salted or Seasoned Soy Curd

To add flavor to soy curd, place the cake in cold, salted water to cover (one teaspoon salt to two cups water) or cold beef or chicken broth (canned, homemade or water in which a bouillon cube or broth granules have been dissolved). Slowly bring the liquid to the simmering point and simmer twenty minutes, until the curd floats. Remove from heat and let set in the liquid until cold.

Soy Sprouts

Soybean sprouts are an easily prepared, highly nutritious fresh vegetable. They can be steamed, fried, creamed, used in salads or added to soups, stews or casseroles.

Used fresh, soy sprouts add a crisp texture to a vegetable dish. They also may be dried or roasted, then crumbled or chopped and used as a topping for casseroles and main dishes, in meat loaves and imitation meat dishes or to add crunch to cookie doughs.

Soy sprouts are rich in protein and the vitamins and minerals for which the soybean is valued. In addition, vitamin C is developed during the sprouting process. Unlike most vegetables, the amount of vitamin C continues to increase for the first few days of storage in the

refrigerator. Two-inch sprouts which have been refrigerated two days have as much vitamin C as an equal amount of citrus fruit.

There's no doubt that soy sprouts are good for you. They're also a convenient fresh vegetable that can be grown as needed and an economical source of vitamins, minerals, and protein. They're excellent for dieters—only sixty-five calories per cup. What's more, they're easy to grow. Here's how:

Growing Soy Sprouts

The first thing to learn about soy sprouts is that it's important to start small. Remember that one pound of soy sprouts will serve six to eight people and that one pound of soybeans will make six pounds of soy sprouts—enough to serve thirty-five to forty people.

No fancy equipment is needed. You can buy special sprouting equipment, of course, but you don't need to. Almost any receptacle large enough to hold the finished sprouts will do. Right in your kitchen you probably have containers made of plastic, glass, china, enamel, or unglazed pottery which are fine.

Look around. You'll find beanpots, crocks, canisters, coffee pots, plastic freezer boxes, one-quart or larger glass canning jars. It's best to have a wide-mouth container for easy rinsing and a diameter large enough so the seeds can be spread out in a thin layer.

Once you've settled on a container, gather up the other things you'll need—a measuring cup, some cheesecloth, plenty of paper towels, and a large mesh strainer.

Other than that, you'll need the proper amount of moisture, enough to keep the seeds moist, but not standing in water; the right temperature, 70 to 80° F.; good circulation of air, (always keep one-third of the container empty for air space) and some fresh, viable, untreated soybeans.

METHOD I

Start with one-third cup of dried soybeans, enough to make two cups of sprouts. It's a good idea to start one evening after dinner.

Put the beans in the strainer and rinse well in cool water. Pour the washed beans into the container you've chosen and cover with one cup of cool water. Put it in a warm (70° to 80° F.) place overnight.

Next day, skim off any seeds floating on top or any that are broken or not swelling. Pour off the water. (It's good for houseplants.) Rinse beans again in the strainer. Drain and spread the wet seeds in the container and fasten a double thickness of cheesecloth over the opening with a string or rubber band.

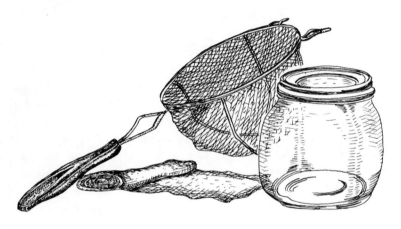

Leave the container in a warm, dark place convenient to the kitchen sink and where you won't forget it. Turn it over so the wet beans rest on the cheesecloth. Be sure to keep the jar in the dark or the sprouts will be tough. If necessary, it may be set on the kitchen counter with a towel draped over the container to keep the sprouts in the dark.

Every four hours during the day, rinse the beans well, then drain completely. Stir the sprouting beans to keep them separated and remove any hulls that have fallen off or beans that are rotting.

Repeat the rinsing process each day for four or five days, until sprouts are two-and-a-half to three inches long. Rinse one last time, drain well and label the container (with masking or adhesive tape) with the date. Refrigerate two or three days before eating, but no more than five or six days. If you like your sprouts green, set the sprout jar in the sun for an hour or two before refrigerating.

Soy sprouts may be steamed five or ten minutes in one-half cup water and served as is or they may be used in many of the dishes in the recipe section.

METHOD II

Use a clean, unused clay flower pot—the kind with a drainage hole in the bottom—for a sprouting container. Line the bottom of the pot with a double thickness of cheesecloth. Pour the soaked beans onto the cheesecloth and cover with a damp cloth. Place the pot on a drainboard, drape a towel over the top to keep it dark and pour one cup water into the pot three times a day. In three to five days, the sprouts will be ready.

METHOD III

Dip a clean, open weave dish towel in cool water. Wring out and spread the towel on a counter top. Sprinkle with one-third cup dried soybean seeds. Beginning at one of the small ends, carefully roll up the towel loosely to make a long roll. Keep in a cool, dark place and each day sprinkle the towel with cool water to keep it damp. In four to five days, unroll the towel and pick off the sprouts.

METHOD IV

Soak soybean seed overnight in cool water to cover. Fill a wooden sprouting box (about 10 x 12 x 6 inches) half full of potting soil or peat moss. Sprinkle soybean seeds over the top and cover lightly with more soil. Put a closely fitting lid on the box and leave undisturbed in a cool place four or five days. There is no need to water the soil; the moisture in the soil and the damp seed are enough for germination.

Dried Soy Sprouts

Dried soy sprouts make a delicious food concentrate to sprinkle into or on almost any food. And, since sprouts are best eaten within three or four days after sprouting, drying them is a good way to make use of any leftovers.

To dry soy sprouts, spread them thinly on a cookie sheet and place the sheet in a well-ventilated place, not in direct sunlight. Stir every hour or so. On a dry, hot summer day, they should dry thoroughly in one day. In rainy weather you may want to slide the tray into a 150° F. oven or into a homemade or commercial food drier.

As the sprouts dry, their flavor changes. When thoroughly dry, they should have a delicious, nut-like flavor. They may be chopped or ground and stored in tightly covered glass containers.

Use dried soy sprouts to add nutrition and good flavor to meat loaves and casseroles, as a topping for vegetables, breakfast cererals, and desserts, in cookie batters and bread doughs, in a sandwich spread or as a snack right from the jar. They may be used as is, without soaking.

For a delicious, nutritious breakfast food, finely chop one cup of dried soy sprouts and add one-fourth cup of seedless raisins, one-fourth cup of chopped soy nuts, one-half cup of uncooked rolled oats, and two tablespoons of honey. Mix well and spread on a cookie sheet to toast. Heat in a 250° F. oven until golden brown and crunchy. Serve as is or with milk.

Dried soy sprouts also may be added to tossed salads, stirred into soups, served on or in spaghetti sauce, or added to scrambled eggs.

Soybean "Meats"

If you have never tasted soyloaf, you're in for a surprise. Properly flavored and well prepared, soyloaf not only is as protein-rich as animal meats, it also has a flavor, texture, and color much like meat. With a little imagination, soyloaf can feed your family well at less than one-fourth the cost of meat.

The basic soyloaf requires a little practice and some time to prepare at first. However, the good nutrition and the savings make it well worth the trouble. To save time and energy—both yours and that used by your kitchen range—it is sometimes possible to double the recipe and freeze one half for later.

Since soybeans have a rather bland flavor, the meaty flavor of soyloaf depends on the seasonings used in these recipes. Many flavorings are available commercially which may be used to give the soyloaf whatever flavor you wish. The addition of any of the following may be used according to your preference:

• Dry broth powders or granules in beef, chicken, or vegetable flavors. These usually are all-vegetable products and are available in health-food stores.

• Beef or chicken bouillon cubes or granules. These may have some animal products in them and are available in grocery stores and supermarkets.

• Canned or homemade beef or chicken broth or concentrate. May be used in place of water called for in the recipe.

• Onion power

• Garlic powder

• Liquid smoke (for a ham or bologna flavor)

• Hickory smoked salt (for a ham or bologna flavor)

• Sage or thyme (for a sausage flavor)

Basics You'll Need for Soyloaves

Adapt the following recipes to suit your taste by varying the seasoning or by using any others you prefer. The amounts listed are light and are intended only as guidelines. After trying them once you may want to use more beef flavoring or to add a touch of garlic or onion. Only by experimenting will you discover the just-right seasoning for your taste.

To make soyloaf, cooked soybeans in some form are mixed with oil, water, salt, flavoring, and flour and poured into molds, then cooked at 15 pounds pressure for 1½ hours. During the cooking process, the mixture develops the firm texture and brown color of meat.

The cooked soyloaf is cooled, removed from the mold, and sliced. It may be served as is, hot or cold, or it may be used in one of the later recipes for soyloaf dishes.

Here is a variety of soyloaf recipes to start you off:

BASIC SOYLOAF

½ cup dried soybeans
4 cups cold water
1 cup soy flour (page 29)
1 cup wheat flour

¼ cup cooking or salad oil
1 tablespoon dry broth powder
  or granules (beef or chicken
  flavor)
1½ teaspoons salt

Put soybeans and water through blender until soybeans are well ground but not liquified. Pour into a saucepan and cook, covered, over low heat 45 minutes. Drain, reserving cooking liquid. To cooked, ground soybeans add soy flour, wheat flour, oil, meat flavoring, and salt. Add enough cooking liquid to make a medium-thin batter (about one cup) and blend well in blender. Pour into two oiled, clean soup cans and cover tightly with aluminum foil. Tie on with string or rubber bands. Cook at 15 pounds pressure 1½ hours. Let cool 10 minutes in cans, until firm, then force from cans by removing the bottoms of cans and pushing the loaf through. Makes about 1½ pounds of soyloaf.

BEEF-FLAVORED SOYLOAF

2 cups cooked, dried
  soybeans, drained
2 teaspoons salt
4 tablespoons unsalted beef
  broth powder or granules
1 cup soybean cooking liquid

½ cup soy flour (page 29)
½ cup wheat flour
1 cup soy nut butter (page 100)
½ cup fresh or canned mushrooms, sliced

Using a pastry blender or potato masher, coarsely chop cooked soybeans. Dissolve salt and broth powder in bean cooking liquid. Combine soy flour and wheat flour and spread on a dry cookie sheet. Bake in 250° oven until lightly toasted. Mix all ingredients well and pour into greased one-pound coffee can and cook 1½ hours in a pressure pan at 15 pounds pressure. Makes 1½ pounds soyloaf.

BOLOGNA SOYLOAF

2 cloves garlic, chopped
1 cup tomato juice
2 cups roasted soy flour
  (page 49)
1 cup wheat flour

1 teaspoon brown sugar
1 tablespoon salt
1 teaspoon onion juice
1½ teaspoons liquid smoke
  flavoring

Add chopped garlic to tomato juice and let set overnight or several hours, tightly covered, in the refrigerator. Strain, reserving the juice and discarding the garlic. Combine juice with remaining ingredients, blending well. Pour into two No. 2 size vegetable cans which have been well greased. Cover with a piece of aluminum foil and fasten with a piece of string or a rubber band. Cook 1½ hours at 15 pounds pressure in a pressure pan, following manufacturer's directions. Cool 10 minutes, then invert can and cut off bottom with a can opener. Force out loaf by pushing bottom end through. Chill. Slice thinly for use in sandwiches or use in later recipes.

CHICKEN SOYLOAF

1 cup dried soybeans
1 cup cold water
¼ cup cooking or salad oil
1 cup wheat flour

2 teaspoons salt
2 tablespoons chicken broth
  powder or granules (unsalted)

Put soybeans and water in blender and run at high speed until soybeans are finely chopped, but some lumps remain. Combine the mixture with oil, flour, salt, and broth powder. (If salted broth powder is used, omit salt in recipe.) Run through blender on low speed until well blended. Pour into two clean, well-greased soup cans. Cover with aluminum foil and fasten tightly with string or rubber bands. Cook 1½ hours in pressure pan at 15 pounds pressure, following manufacturer's directions. Let pressure drop of its own accord. Cool 10 minutes in cans until solid. While still warm, invert can and remove bottom and push to force loaf out of cans. Use as a main dish, for sandwiches or as an ingredient for the recipes to follow. Makes about 1½ pounds of soyloaf.

HAM SOYLOAF

2 cups roasted soy flour (page 49)   1 cup water
1 cup wheat flour   ¼ cup cooking or salad oil
2 tablespoons brown sugar   1½ teaspoons hickory smoked
                                             salt

Combine all ingredients, blending well in blender. Pour into two well-oiled, clean soup cans, cover with aluminum foil and fasten with string or rubber bands. Cook 1½ hours at 15 pounds pressure in pressure pan. Let pressure drop of its own accord. Cool 10 minutes, then remove from cans by cutting off bottom end, then forcing loaf out through the top. Slice and serve hot, chill, or use in the recipes to follow.

Roasted Soybeans

Soybeans may be roasted to improve the flavor and texture and to change the color for some food uses. Roasting also serves as a form of cooking to destroy the anti-trypsin factor which makes the uncooked soybean protein difficult to digest.

Soybeans may be roasted in any form—whole, cracked, ground, milled, or sprouted. Roasting removes the "beany" flavor and gives the soy product a nut-like taste which improves many dishes.

Whole roasted soybeans may be eaten as a snack, like peanuts, or used in any way peanuts are used. Grind them and add to casseroles or as dessert toppings, in candies or cookie batter, or in any recipe using ground nuts. Roasted soy flour adds a meat-like flavor to meat substitutes, meat extenders, and gravy.

There are two methods of roasting soybeans. In the dry roasting process, soybeans are toasted in the oven at a low temperature. Dry roasting may be used for soybeans in any form and is the only method possible for soy grits or soy flour.

Oil roasting, or deep-fat frying, is the preferred method of making soy nuts and roasted soy sprouts if the additional fat in the diet is no problem. Whole soybeans prepared in this way taste much like roasted peanuts.

Soy grits and soy flour may be dry roasted in any order. The soybeans may be roasted whole, then ground or milled, or the soaked, dried soybeans may be ground or milled, then roasted. Instructions for both methods are included.

Roasting Whole Soybeans

OVEN-ROASTED
SOY NUTS

1 cup dried soybeans
2 cups cold water (about 50
  degrees)

1 tablespoon cooking or salad
  oil
Salt

Pour cold water over soybeans and let soak 1½ hours. Drain off the water and pat the beans dry with towels. Spread out thinly on an ungreased cookie sheet and roast in a 350° oven about 30 minutes, until lightly browned, stirring every 10 minutes. Remove from oven, drizzle with oil, and stir to coat. Sprinkle with salt to taste.

OIL-ROASTED
SOY NUTS

1 cup dried soybeans
2 cups cool water (about 70°)

Hot oil for frying
Salt

Soak soybeans two hours in cool water to cover. Drain and spread on paper towels, patting to remove as much water as possible. Add, a few at a time, to oil at least one inch deep which has been heated to 375°. Add carefully, for oil will bubble up. Fry three or four minutes, until crisp and lightly browned. Remove with slotted spoon. Drain on paper toweling and sprinkle with salt.

Either dry roasted or oil roasted soy nuts may be eaten as is, mixed with salted peanuts, or flavored according to the recipes to follow.

Or they may be coarsely ground in a blender or food grinder and used as a nut-like topping to add interest and protein to casseroles, vegetable dishes, and desserts. Whole and ground soy nuts are used as ingredients for several dishes in the recipe section.

To mill a rich, nut-flavored flour from dry roasted soybeans, grind soy nuts very coarsely and remove the hulls by winnowing. To winnow, slowly pour the coarsely ground soy nuts in front of a fan. The nuts will drop to a tray or pan and the hulls will blow to the side. Two or three passes may be needed to remove all the hulls.

Now the roasted soybeans may be milled into roasted soy flour, using a hand-cranked grinder or electric flour mill according to the manufacturer's instructions.

This flour may be used to add flavor and nutrition to meatloaves and breads.

DRY-ROASTED SOY GRITS

1 cup dried soybeans   3 cups cold water

Soak soybeans overnight in cold water. Refrigerate to keep from fermenting. Drain and spread on a towel to dry. Run through coarse blade of food grinder or through blender quickly. Rinse in water to cover and skim loose hulls off the top. Drain again on dry towel and spread thinly on ungreased cookie sheet. Dry roast in a 300° oven 30 minutes, until lightly roasted, stirring occasionally.

DRY-ROASTED SOY FLOUR

1 cup dried soybeans   3 cups cold water

Soak soybeans overnight in water in refrigerator. Drain. Crack beans by running through coarse blade of food grinder or quickly through a blender. Rinse in cold water to cover and skim off hulls that rise to the top. Drain thoroughly and spread cracked soybeans out on ungreased cookie sheet. Set in 300° oven until dry, stirring occasionally.

Grind to a fine powder in hand grinder or electric flour mill. Spread soy flour thinly over cookie sheet and dry roast in 300° oven about 20 minutes, until lightly toasted. Watch carefully and stir occasionally to toast evenly.

Roasted soy flour gives a pleasant, nutty taste to foods and is ready to use as is. The roasting process destroys the anti-trypsin factor and eliminates the need for cooking the raw beans before milling.

DRY-ROASTED SOY SPROUTS

2 cups soy sprouts (page 38)   Salt

Rinse soy sprouts and spread on absorbent towel. Pat as dry as possible. Spread thinly on greased cookie sheet and roast in a 350° oven about 30 minutes, stirring occasionally, until sprouts are golden. Salt to taste.

OIL-ROASTED SOY SPROUTS

2 cups soy sprouts (page 38)   Salt
Oil for deep frying

Rinse and drain soy sprouts. Spread a thin layer on absorbent towels and pat dry. Heat at least one inch of oil to 350° and deep fry sprouts, a few at a time, until golden. Salt to taste.

Roasted soy sprouts may be used in any of the ways listed for dried soy sprouts, page 41. They also may be used to make a spread that tastes much like peanut butter. Just put them through the blender until smooth, add salt, and blend in cooking or salad oil until the spread is the consistency of peanut butter.

Fermented Soybean Cakes

This traditional Oriental form of soybean "meat" is made by introducing the fermenting culture to cooked, dried soybeans, then keeping it under the ideal, warm, moist conditions which allow the culture to develop.

The process is not mysterious, but is similar to the making of yogurt or cultured buttermilk. The resulting product—like yogurt or buttermilk—is more digestible than the original food because it has been pre-digested by the enzyme action of the culture. The fermentation process not only digests the beans and tenderizes them, it also binds the beans together in a firm, flat loaf which can be cut into patties. It then requires only a few minutes to cook, and is usually fried. The flavor is mild. Raw, it tastes like cheese. Fried, it has a meaty flavor, something like veal.

To make one pound of fermented soybean cake, soak 1 cup dried soybeans overnight in 3 cups cold water. In the morning, put the beans and water through the coarse blade of a food grinder or quickly through the blender. Beans should remain in fairly large pieces.

Do not drain. Add water if necessary to cover ground beans. Heat to simmering and cook 40 minutes, stirring frequently and skimming away any hulls that float to the top. Cool 10 minutes. Add 1 tablespoon vinegar and mix well, then drain through a colander, reserving the liquid for soup stock. (Any vinegar in the liquid will be good for the soup, leaching the calcium from soup bones into the stock, then evaporating as the soup cooks.)

Spread the drained soybeans out on paper towels and cool. When lukewarm, pour them into a mixing bowl and stir well with the fingers to make sure all the beans are cooled. The fermenting starter is like yeast—it can be killed by heat. Put one teaspoon fermenting starter into a clean salt shaker and sprinkle the starter evenly over the lukewarm soybeans, mixing well to coat all the beans.

Spread the coated soybeans in a layer ½-inch deep in the bottom of a square cake pan or flat-bottomed casserole. Cover the top with aluminum foil and punch holes with a fork every two inches or so across the top to allow enough air to let the mold grow but not enough to let it dry out.

Set the pan in a warm, draft-free place where the temperature can be kept between 80° and 90°. From 85° to 88° is ideal if you can control it that closely. An electric oven with a low setting or a gas oven with a pilot light and a pan of hot water will work well. So will a styrofoam picnic cooler with a light bulb or a heating pad inside. Experiment a little with a thermometer until you find a spot the right temperature.

Once you have found it, leave the pan with the mixture undisturbed about twenty hours, then check it. If the fermented cake smells fresh and clean—something like mushrooms—and the soybeans are bound together in a compact mass with the beans barely visible, the soybean cake is ready.

If there are holes here and there and the bean cake is crumbly, put it back in its warm place another four to eight hours. The exact time depends on the temperature.

If there is a strong, ammonia smell, the fermentation has gone too far. If ever, at any stage along the way, the beans smell unpleasant or spoiled, discard them and start over. It happens with wine or cheese or bread, too, so count it as part of the learning process.

Once the fermented soybean cake has reached the just-right stage, cut it into strips or squares and season to taste with salt. The pieces may be dipped in a solution of 1 cup water to ½ teaspoon salt or sprinkled with salt, garlic salt, or onion salt.

Heat vegetable oil in a skillet to 350° and fry cakes on one side until golden, then turn and brown on the other side. Serve as a main dish, on sandwich buns or in casseroles.

Fermented soybean cakes do not keep more than a day or two in the refrigerator, but can be frozen or dried.

You can obtain fermenting starter (usually called by its Japanese name, *Tempus Starter*) in health food stores and specialty sections of large supermarkets. It may be ordered by mail from either of the following:

THE FARM, Summertown, TN. 38483

DR. HWA WANG, U.S. Department of Agriculture, Northern Regional Research Laboratory, Peoria, IL 61604.

Soy Beverages

Soy milk can be used in any beverage in which cow's or goat's milk might be used. Since it can be made as rich as desired by adding soy flour, soy milk is excellent for milk shakes, eggnog, hot chocolate, and yogurt. In addition, delicious beverages can be made by liquifying soy sprouts.

Following are some suggested recipes for using soy products in beverages. Use them as guidelines for creating your own.

SOY BUTTERMILK

½ cup cultured buttermilk,   4 cups soy milk (page 33)
    warmed to room temperature

Stir buttermilk into soy milk, blending well. Cover and let stand at room temperature (70° to 80° F.) or in yogurt maker for two hours. Shake well, then return to a warm place until clabbered (about six hours). Refrigerate until used. Always save the last one-half cup as a starter for the next batch.

SOY EGGNOG

2 eggs, separated
4 teaspoons soy flour (page 29)
2 cups soy milk (page 33)

½ teaspoon vanilla
8 teaspoons sugar
Dash nutmeg

Beat egg yolks until light. Blend in soy flour, one teaspoon at a time. Gradually add soy milk, beating after each addition. Add vanilla. Beat egg whites to soft peaks. Gradually add sugar, beating constantly. Fold into milk mixture and pour into chilled glasses. Top with a sprinkling of nutmeg. Serves four.

SOY MILK SHAKE

1 egg
2 tablespoons honey

1 tablespoon lemon juice
1 cup soy milk (page 33)

Beat egg until light and creamy. Add honey, lemon juice and soy milk. Beat three minutes. Makes one serving.

BANANA-PINEAPPLE SOY SHAKE

2 ripe bananas
2 cups pineapple juice

2 cups soy milk (page 33)

Combine all ingredients. Blend two minutes. Makes four servings.

CHOCOLATE SOY SHAKE

1 cup soy milk (page 33)
¼ cup soy flour (page 29)
3 tablespoons chocolate syrup

½ teaspoon vanilla
1 cup soy ice cream (page 87)

Blend all ingredients in shaker or blender. Serves two.

SOY YOGURT

3 cups soy milk (page 33)   ½ cup yogurt (homemade or
                                       commercial, with live culture)

Warm soy milk to room temperature. Stir in yogurt and blend well.
Keep warm in yogurt maker, thermos bottle or in warm water over
pilot of kitchen gas range. Let set six to eight hours, until firm and as
tart as desired. Save last one-half cup to start the next batch.

STRAWBERRY YOGURT

1 cup fresh or frozen   1 cup soy yogurt (above)
  strawberries, well drained

Carefully fold strawberries into yogurt. Makes two cups.

SOY SPROUT MILK

1 cup soy sprouts (page 38)   2 tablespoons honey
4 cups warm water

Blend soy sprouts, water, and honey in blender five minutes at high
speed. Cook over medium heat 10 minutes, stirring constantly. Strain.
Let cool and use strained liquid as a beverage, sauce base, soup stock,
or liquid for breads. Use sieved residue in meat loaves, meat stuffing,
casseroles, or vegetable loaves.

SOY SPROUT COCKTAIL

1 cup soy sprouts (page 38)   1 tablespoon honey
2 cups fruit juice (pineapple,
  apple, berry, or orange)

Blend all ingredients three minutes in blender or shaker. Strain. Serve
over crushed ice. Makes 2½ cups.

SOY COFFEE SUBSTITUTE

To make a really delicious hot beverage, dry roast whole soybeans without soaking in a 200° oven four to five hours, until lightly browned all the way through. Watch carefully and stir frequently the last hour to keep from burning. Cool, then grind in a blender, food grinder, or coffee grinder, adjusting the grind from fine to coarse, as preferred. The beans may be roasted and stored in air-tight containers for a week or two at a time, but for best flavor, grind only enough for one batch at a time. Like coffee, it is most flavorful when freshly ground.

Brew as you would coffee, in a coffee pot, coffee maker or in a saucepan. It may be used alone or mixed half and half with coffee.

Soy Breads

Soy flour adds to the shelf life of yeast breads and quick breads which keep better because the products are more moist. It also improves the quality of deep-fried doughnuts, absorbing less of the fat in which they are fried than do doughnuts made of all wheat flour.

Breads made with soy flour are slightly heavier, with a denser texture. They also brown more easily, so they should be baked at a slightly lower temperature.

In the following recipes, when the term "wheat flour" is used, either whole wheat or white all-purpose wheat flour may be used, according to preference. When whole wheat or white flour is more appropriate to the recipe, the type is specified.

The following is only a sampling of the wide variety of breads in which soy products may be used.

BANANA BREAD

1 tablespoon vinegar
½ cup soy milk (page 33)
1¾ cups wheat flour
¼ cup soy flour (page 29)
1 teaspoon soda
1 teaspoon salt

½ cup roasted soy grits (page 00)
½ cup solid vegetable shortening
1 cup sugar
2 eggs
1 cup mashed bananas (2 or 3
   bananas)

Add vinegar to soy milk. Set aside to curdle. Combine wheat flour, soy flour, soda, salt, and soy grits in a small bowl. In a larger mixing bowl, cream shortening and sugar. Add eggs, one at a time, then mashed bananas, beating well after each addition. In small amounts, add dry ingredients alternately with curdled soy milk, beating well by hand or with an electric mixer. Pour into greased 9 × 5-inch loaf pan. Bake in 325° oven 60 to 70 minutes, or until a toothpick inserted comes out clean. Makes one loaf.

SOY BREAD

1 cup soy milk (page 33)
2 tablespoons sugar
1 teaspoon salt
2 tablespoons cooking or salad
   oil
1 cup water

1 package yeast or 1 tablespoon
   dry yeast
¼ cup lukewarm water
1½ cups soy flour (page 29)
4½ cups wheat flour

Heat soy milk to just under boiling. Pour into large mixing bowl and add sugar, salt, and oil. Stir well, then add water and let set until lukewarm, stirring occasionally. In a cup, combine yeast and ¼ cup lukewarm water. Let set five minutes. Add to lukewarm soy milk mixture and mix well. Add both flours, one cup at a time, mixing until dough is formed. Knead on a lightly-floured kneading board until smooth and no longer sticky. Place in a greased, warm bowl, cover with a damp cloth and let rise 1½ hours in a warm, draft-free place, until doubled in bulk. Punch down. Let rise again, about one hour, until doubled in bulk. Divide into two parts, cover and let rise 15 minutes. Shape into two loaves and place in greased loaf pans. Let rise one hour, or until center of loaf is slightly higher than sides of pan. Bake in 325° oven one hour, or until golden brown and loaf sounds hollow when thumped with finger. Makes two loaves.

SOY BISCUITS

1¾ cups wheat flour
¼ cup soy flour (page 29)
3 teaspoons baking powder
1 teaspoon salt

4 tablespoons solid vegetable
  shortening
¾ cup soy milk (page 33)

Sift together dry ingredients. Blend in shortening with a pastry blender or two knives until the consistency of coarse meal. Add soy milk, mixing lightly with a fork until dough leaves the sides of the bowl. Knead five or six times, then pat out dough to one-half inch thickness and cut with a biscuit cutter. Bake in an ungreased pan 12 to 15 minutes in a 400° oven. Makes 15 to 18 biscuits.

LOW-CHOLESTEROL
SOY BISCUITS

1½ cups wheat flour
½ cup soy flour (page 29)
3 teaspoons baking powder

1 teaspoon salt
⅓ cup safflower or corn oil
⅔ cup soy milk (page 33)

Sift dry ingredients together. Pour oil and soy milk together in a cup. Stir well. Pour, all at once, into the flour mixture. Stir with a fork until dough leaves the sides of the bowl. Knead five or six times. Pat one-half inch thick and cut with a biscuit cutter. Bake 10 to 12 minutes on an ungreased cookie sheet in a 400° oven. Makes 15 to 18 biscuits.

SOY GRITS BISCUITS

1½ cups wheat flour
1 teaspoon salt
3 teaspoons baking powder
½ cup roasted soy grits (page 00)

2 tablespoons solid vegetable
  shortening
⅔ cup soy milk (page 33)
2 tablespoons molasses

Combine flour, salt, baking powder, and soy grits. Cut in shortening
with pastry blender or two knives to a fine crumb consistency. Combine
soy milk and molasses and add. Beat well to blend. Drop by spoonfuls
onto a greased baking sheet. Bake in 400° oven 15 to 18 minutes, until
lightly browned. Makes 12 biscuits.

SOY NUT BUTTER BREAD

1¾ cups wheat flour
¼ cup roasted soy flour (page 00)
4 teaspoons baking powder
1 teaspoon salt

⅓ cup sugar
1 cup soy milk (page 33)
⅔ cup soy nut butter (page 100)

Sift all dry ingredients together. In a separate bowl, mix soy milk and
soy nut butter. Add to dry ingredients and beat until blended. Pour
into greased 5 × 8-inch loaf pan and bake one hour in 325° oven.
Makes one loaf.

SOY CORNBREAD

¾ cup yellow cornmeal
1 cup wheat flour
¼ cup soy flour (page 29)
½ cup sugar
½ teaspoon salt

3 teaspoons baking powder
1 egg
¾ cup soy milk (page 33)
¼ cup cooking or salad oil

Combine dry ingredients in mixing bowl. In another bowl, combine
egg, soy milk and oil. Add to dry ingredients and stir just enough to
blend. Pour into greased 8-inch square baking pan and bake 25 minutes
in a 400° oven. Cut into squares to serve. Makes nine large servings.

SOY POTATO BREAD

1 medium potato
6 cups water
2 cakes yeast or 2 tablespoons
  dry yeast
5 tablespoons honey

4 teaspoons salt
4 tablespoons cooking or salad
  oil
7 cups wheat flour
3 cups soy flour (page 29)

Peel and chop potato and cook in 6 cups water until tender. Run potato and cooking water through ricer or blender until smooth. Cool to lukewarm, then stir in yeast, honey, salt, and oil. Combine flours and add, one cup at a time. Knead well. Let rise in a warm place until doubled in bulk. Punch down, then let rise again. Punch down and shape into loaves. Let rise in a warm place until centers of the loaves are slightly higher than the sides of the pans. Bake one hour in 325° oven. Makes four loaves.

SOY—MIXED WHEAT BREAD

1½ cups soy milk (page 33)
¼ cup molasses
1½ teaspoons salt
2 tablespoons cooking or salad
  oil

2 cakes yeast or 2 tablespoons
  dry yeast
1 cup soy flour (page 29)
1½ cups whole wheat flour
1½ cups all-purpose white
  wheat flour

Heat soy milk to lukewarm. Add molasses, salt, and oil. Stir well, then add yeast. Let set until yeast begins to bubble, then add soy flour, whole wheat flour, and enough white flour to make a dough which is easily handled. Knead 10 minutes, adding more white flour if needed. Place dough in a warm, greased bowl. Cover and let rise until doubled in bulk, about 1½ hours. Punch down and let rise again, about one hour. Punch down and shape into a loaf. Let rise until the center is slightly higher than the sides of the pan. Bake one hour in a 325° oven. Makes one large or two small loaves.

WHOLE WHEAT – SOY CRACKERS

½ cup honey
½ cup cooking or salad oil
1 cup soy milk (page 33)
3½ cups whole wheat flour

½ cup roasted soy flour
  (page 49)
½ teaspoon salt

Blend together honey, oil, and soy milk. Combine dry ingredients and add all at once to liquid. Beat well. Roll out thin on floured board. Cut into two-inch squares or circles and bake on an oiled cookie sheet in a 325° oven 15 to 20 minutes, or until lightly browned. Makes 8 to 9 dozen crackers.

SOY – WHEAT GERM CRACKERS

¾ cup solid vegetable
  shortening, melted and cooled
1¼ cup soy milk (page 33)
1 teaspoon salt
1 cup wheat germ

2 tablespoons honey
½ cup roasted soy flour
  (page 49)
3½ cups whole wheat flour

Mix all ingredients, using electric mixer or beating by hand. Knead five minutes. Roll out very thin and cut into squares. Pierce each cracker several times with a fork. Bake in a 300° oven until lightly browned. Makes 9 or 10 dozen crackers.

BOSTON BROWN SOY BREAD

2 cups soy milk (page 33)
2 tablespoons vinegar
1½ cups soy flour (page 29)
1½ cups whole wheat flour
1 cup yellow cornmeal

½ cup sugar
½ teaspoon salt
2 teaspoons soda
1 cup molasses
1 cup seedless raisins

Mix soy milk and vinegar and set aside to curdle. Meanwhile, in a large bowl, mix soy flour, whole wheat flour, cornmeal, and sugar. In another bowl, stir salt and soda into soy milk-vinegar mixture. Add, all at once, to flour mixture. Add molasses and raisins and beat just until well blended. Divide batter into five well-greased 16-ounce baking cans (or No. 2 cans from canned vegetables). Cover each can with aluminum foil and tie down with string or rubber bands. Place on a rack set in a large kettle. Fill kettle with boiling water to a depth of two inches and adjust heat to keep water to just under boiling. Cover kettle and cook one hour, until bread is firm to the touch. Makes five small, round loaves.

SOY GRIDDLECAKES

2 eggs, well beaten
1½ cups soy milk (page 33)
2 tablespoons cooking or salad oil
1⅔ cups wheat flour

⅓ cup soy flour (page 29)
3 teaspoons baking powder
½ teaspoon salt
1 tablespoon sugar

Combine eggs, soy milk, and oil. Add to dry ingredients. Beat until smooth. Pour by one-fourth cupfuls onto hot, lightly-greased griddle and cook until bubbles on top burst. Flip and cook on other side. Makes 12.

SOY-OAT GRIDDLECAKES

2 cups soy milk (page 33)
1½ cups rolled oats (uncooked)
½ cup wheat flour
⅓ cup soy flour (page 29)

2½ teaspoons baking powder
1 teaspoon salt
2 eggs, separated
⅓ cup cooking or salad oil

Heat soy milk almost to boiling and pour over oats. Allow to cool. Stir together flours, baking powder, and salt. Beat egg yolks and add to oat mixture. Add oil and stir in dry ingredients. Fold in stiffly beaten egg whites. Drop batter by large spoonfuls onto hot, greased griddle. When surface is covered with bubbles, flip and brown on other side. These take longer to bake than other griddlecakes. Makes about 12.

SOUR MILK GRIDDLE CAKES

¼ cup vinegar
1¼ cups soy milk (page 33)
1¾ cups wheat flour
¼ cup soy flour (page 29)
1 teaspoon soda

2 teaspoons baking powder
3 tablespoons sugar
1 teaspoon salt
2 eggs, well beaten
3 tablespoons cooking or salad oil

Add vinegar to soy milk. Set aside 10 minutes to curdle. Combine wheat flour, soy flour, soda, baking powder, sugar, and salt. Stir soured soy milk and add all at once, to dry ingredients. Add eggs, then oil, beating after each addition. Pour, one-half cupful at a time, onto greased hot griddle. Cook until bubbles form, then flip and brown on other side. Makes about 12.

GINGERBREAD

2 cups wheat flour
½ cup soy flour (page 29)
1 teaspoon salt
1 teaspoon baking powder
1 teaspoon powdered ginger
2 teaspoons powdered cinnamon
½ teaspoon powdered cloves

½ cup solid vegetable shortening
½ cup sugar
¾ teaspoon soda
1 cup dark molasses
2 eggs
1 cup hot water

Sift together first seven ingredients. Set aside. In another bowl, cream shortening, sugar, and soda. Add molasses. Beat well. Stir in one-half cup flour mixture and beat in eggs. Add hot water alternately with remaining flour mixture, adding small amounts at a time. Beat one minute. Pour into a well-greased, lightly floured 9 × 9 × 2-inch pan. Bake 45 minutes in a 325° oven.

SOY DOUGHNUTS

3 tablespoons butter
1 cup sugar
2 eggs, well beaten
3¼ cups wheat flour
½ cup soy flour (page 29)

4 teaspoons baking powder
½ teaspoon salt
¾ cup soy milk (page 33)
1 teaspoon vanilla
Deep fat for frying

Cream butter and sugar until light. Stir in beaten eggs. Sift dry ingredients together and add to creamed mixture alternately with the milk. Stir in vanilla. Chill several hours or overnight. Roll out one-third-inch thick on a floured surface. Cut with floured doughnut cutter and fry in deep fat which has been heated to 375°. Brown on one side, then turn and brown on the other. Makes three dozen doughnuts.

SOY-POTATO DOUGHNUTS

1 cup soy milk (page 33)
1 tablespoon vinegar
4 cups wheat flour
½ cup soy flour (page 29)
1 teaspoon salt
1 teaspoon soda
1 teaspoon nutmeg

2 eggs, well beaten
1 cup sugar
2 tablespoons cooking or salad
  oil
1 cup cooked potato, drained
  and mashed
Deep fat or oil for frying

Combine soy milk and vinegar. Set aside to curdle about 10 minutes. Sift together wheat flour, soy flour, salt, soda, and nutmeg. In another bowl, beat eggs and sugar until light. Add oil, potato, and soy milk mixture, beating until smooth. Stir in flour mixture. Chill several hours. Roll on a floured surface to one-half inch thickness. Cut with a doughnut cutter and fry in deep, hot (375°) fat or oil. Drain. Makes three dozen doughnuts.

SOY MUFFINS

1 cup wheat flour
1 cup finely ground soy nuts
  (page 47)
3 teaspoons baking powder
¾ teaspoon salt

2 tablespoons sugar
2 eggs, well beaten
¾ cup soy milk (page 33)
1 tablespoon cooking or salad
  oil

Combine wheat flour, soy nuts, baking powder, salt, and sugar. In another bowl, combine eggs, soy milk, and oil. Pour into dry ingredients and stir just enough to moisten. Do not beat. Fill greased muffin tins two-thirds full. Bake 20 to 25 minutes in 375° oven. Makes 12.

BLUEBERRY SOY MUFFINS

1¾ cups wheat flour
¼ cup soy flour (page 29)
1 teaspoon salt
¼ teaspoon soda
2¼ teaspoons baking powder

1 tablespoon vinegar
1 cup soy milk (page 33)
1 egg, beaten slightly
¼ cup cooking or salad oil
½ cup blueberries

Mix dry ingredients in a large bowl. In another bowl, add vinegar to soy milk and let set 10 minutes. Stir egg and oil into milk mixture and add all at once to dry ingredients, mixing just enough to moisten. Fold in well-drained blueberries. Fill greased muffin tins two-thirds full. Bake 25 minutes in 375° oven. Makes 12.

SOY BRAN MUFFINS

¾ cup wheat flour
¼ cup soy flour (page 29)
2½ teaspoons baking powder
½ cup sugar
¾ teaspoon salt
¼ cup solid vegetable shortening

1 cup whole bran cereal
½ cup soy nuts (page 47),
  coarsely ground
1 egg
¾ cup soy milk (page 33)

Combine wheat flour, soy flour, baking powder, sugar, and salt in mixing bowl. Cut in shortening with pastry blender or two knives.

Add bran and ground soy nuts and mix well. In a small bowl, combine egg and soy milk. Add, all at once, to dry ingredients, stirring until flour is just moistened. Fill greased muffin tins two-thirds full and bake 20 to 25 minutes in 375° oven. Makes 12.

SOY YEAST MUFFINS

1 cake yeast or 1 tablespoon dry yeast
¼ cup lukewarm water
¾ teaspoon salt
4 tablespoons honey
2 tablespoons cooking or salad oil

1 cup soy milk (page 33), heated to lukewarm
2½ cups wheat flour
1 egg, well beaten
1 cup soy flour (page 29)
1 egg white, beaten slightly
2 tablespoons soy nuts (page 47), chopped

Soften yeast in lukewarm water. Add salt, honey, and oil to warm soy milk. Blend well. Add yeast and 1½ cups wheat flour. Beat well. Add egg, remaining wheat flour, and soy flour. Mix well. Cover and let rise in a warm place until doubled in bulk, about 30 minutes. Shape into 24 small balls and drop into well-greased muffin tins. Cover and let rise until light and doubled in bulk. Bake in 375° oven about 15 minutes, until lightly browned. Brush tops with egg white and sprinkle with chopped soy nuts. Return to oven and bake 10 minutes longer, until browned. Makes 24 muffins.

SOY NOODLES

1 egg, well beaten
½ teaspoon salt

½ cup soy flour (page 29)
1½ cups wheat flour

Combine egg and salt with a fork. Stir in soy flour and ½ cup wheat flour. Beat well with a fork. Gradually add enough wheat flour to make a very stiff dough, kneading the last of the flour in by hand. Roll out very thin on a well-floured surface and let set 30 minutes, then dust with flour and roll up tightly. Cut into very thin slices and spread noodles out on floured surface to dry. Makes about one pound of noodles.

SOY NUT BREAD

1 cup soy flour (page 29)
1½ cups wheat flour
2 tablespoons sugar
3 teaspoons baking powder
1 teaspoon salt
½ teaspoon powdered cinnamon

1 cup soy nuts (page 47), coarsely
  ground
2 eggs, well beaten
1 cup soy milk (page 33)
4 tablespoons cooking or salad
  oil

Sift together soy flour, wheat flour, sugar, baking powder, salt, and cinnamon. Stir in ground soy nuts. In another bowl, combine eggs, soy milk, and oil. Add, all at once, to dry mixture. Stir with a fork until dough pulls away from the sides of the bowl. Knead quickly on lightly-floured surface, then pat out in a 9 × 9-inch square cake pan. Bake one hour in a 325° oven. Makes nine large servings.

SOY OATMEAL BREAD

2 tablespoons vinegar
2 cups soy milk (page 33)
2 eggs
1 cup sugar
⅔ cup molasses
2½ cups wheat flour
½ cup soy flour (page 29)

1 teaspoon salt
1 teaspoon baking powder
2 teaspoons soda
1½ cups rolled oats
½ cup roasted soy grits
  (page 48)
1½ cups seedless raisins

Add vinegar to soy milk. Let set 10 minutes. Meanwhile, beat eggs until light and add sugar gradually, beating until fluffy. Add soured soy milk and molasses, mixing well. Sift together wheat flour, soy flour, salt, baking powder, and soda. Add to creamed mixture. Add rolled oats, soy grits, and raisins, stirring to combine well. Grease an 8 × 5-inch loaf pan and line with waxed paper. Pour batter in and bake one hour in a 325° oven. Store overnight before slicing.

SPOONBREAD

⅔ cup yellow cornmeal
⅓ cup soy grits (page 27)
1 teaspoon salt

3 cups soy milk (page 33)
2 tablespoons cooking or salad oil
2 eggs

Combine cornmeal, grits, and salt with 2 cups soy milk. Bring to a boil, reduce heat, and cook over low heat five minutes, or until thick, stirring constantly to keep from sticking. Add remaining 1 cup soy milk, oil, and eggs. Beat well. Pour into greased 1½-quart baking dish and bake in 325° oven about 40 minutes, or until a knife inserted in the center comes out clean. Serves four to six.

ALL-GRAIN SPROUT BREAD

(A moist, dark, Pumpernickel-type bread. Grind the grains in the same flour mill you use to make soy flour.)

3 tablespoons honey
2 cakes yeast or 2 tablespoons
   dry yeast
1 cup lukewarm water
3 cups hot water
1½ tablespoons salt
1 cup soy sprouts (page 38)

3 tablespoons cooking or salad
   oil
1 cup soy flour (page 29)
1 cup rye flour
1 cup barley flour
1 cup corn flour
5 cups whole wheat flour

Dissolve 1 tablespoon honey and yeast in 1 cup lukewarm water. Let set 10 minutes. Meanwhile, put 1 cup hot water, remaining 2 tablespoons honey, salt, soy sprouts, and oil in blender. Blend until smooth. Pour into large bowl and add remaining 2 cups hot water. Combine five flours and mix well. Add 5 cups flour mixture to sprout mixture and beat well. Add 1 more cup flour. Cover and let set in a warm place 15 minutes. Add 3 more cups flour, enough to make a soft dough. Knead 10 minutes on floured surface. Shape into four loaves and put into greased 8 × 5-inch bread pans. Let rise 45 minutes, until doubled in bulk. Bake one hour in a 325° oven. Let set one hour before slicing.

SOURDOUGH SOY BREAD

7½ cups whole wheat flour
1½ cups soy flour (page 29)
2 eggs
3 cups sourdough starter
2 cups water

1 tablespoon salt
3 tablespoons cooking or salad oil
1 teaspoon soda
1 cup all-purpose white wheat
  flour

Combine whole wheat and soy flours. In a large mixing bowl, blend eggs, sourdough starter, and water, beating well. Add combined flours and beat 100 strokes. Cover and set in a warm place until doubled in bulk, about four hours. Beat down, add salt, oil, and soda which has been mixed with the white flour. Mix well. Knead 10 minutes on a floured board, adding more white flour, if needed. Form into three loaves and place in greased 8 × 5-inch bread pans. Let rise one hour. Bake in 325° oven about one hour, or until golden brown.

SPROUT BREAD

3 cups soy milk (page 33), heated
  to lukewarm
2 yeast cakes or 2 tablespoons
  dry yeast
1 tablespoon salt
¼ cup honey
3 tablespoons cooking or salad oil

3½ cups all-purpose white
  wheat flour
1 cup ground soy sprouts
  (page 38), drained
1 cup whole soy sprouts (page 38)
2 cups whole wheat flour

Pour 1 cup lukewarm soy milk into large warm bowl. Stir in yeast until dissolved. Add remaining soy milk, salt, honey, and oil. Stir in white flour and beat well. Cover with a damp towel and set in a warm place, until doubled in bulk. Stir down and add ground and whole soy sprouts. Work in two cups whole wheat flour and knead 10 minutes, until smooth and elastic, adding more flour if needed. Place in a clean, oiled, warm bowl and cover with a damp cloth. Let rise in a warm, draft-free place until doubled in bulk. Punch down and knead again. Shape into two loaves and place in greased 8 × 5-inch loaf pans. Bake one hour in a 325° oven. Makes two loaves.

SOY STICKS

1 cup all-purpose white wheat
  flour
½ cup whole wheat flour
½ cup roasted soy grits
  (page 48)
¼ cup sugar

½ teaspoon salt
3 teaspoons baking powder
1 egg, well beaten
¾ cup soy milk (page 33)
¼ cup butter or margarine,
  melted

Combine both flours, soy grits, sugar, salt, and baking powder. Add beaten egg and soy milk. Mix well. Stir in melted butter or margarine. Pour into greased cornstick pans and bake 25 to 30 minutes in 400° oven. Makes 12.

SOY WAFFLES

1¾ cups wheat flour
¼ cup soy flour (page 29)
2 teaspoons sugar
3 teaspoons baking powder
¼ teaspoon salt

3 eggs yolks, well beaten
1½ cups soy milk (page 33)
5 tablespoons cooking or salad oil
3 egg whites, beaten to stiff
  peaks

Combine both flours, sugar, baking powder, and salt. Add, a little at a time, to the egg yolks, alternating with soy milk, and blending well after each addition. Stir in oil. Fold mixture into stiffly beaten egg whites. Bake in waffle iron. Makes six.

FRENCH TOAST WAFFLES

1 egg, well beaten
¼ cup soy milk, (page 33)
2 tablespoons cooking or salad oil

⅛ teaspoon salt
Bread slices

Combine egg, soy milk, oil, and salt. Dip bread slices into liquid, then toast in hot waffle iron.

BUTTERMILK SOY WAFFLES

2 tablespoons vinegar
1½ cups soy milk (page 33)
1½ cups wheat flour
¼ cup soy flour (page 29)
2 teaspoons baking powder

1 teaspoon soda
½ teaspoon salt
3 eggs, separated
½ cup cooking or salad oil

Stir vinegar into soy milk and set aside 10 minutes. In a mixing bowl, combine dry ingredients well. Add soy milk mixture and mix well. Add well beaten egg yolks. Stir in oil and mix lightly. Using an electric mixer or egg beater, beat egg whites until stiff enough to hold peaks. Fold in. Bake in hot waffle iron. Makes eight medium waffles.

SOY-WHEAT PUFFS

1 egg, separated
1 cup soy milk (page 33)
2 tablespoons cooking or salad oil
½ teaspoon salt

½ cup soy flour (page 29)
½ cup whole wheat flour
½ cup all-purpose white wheat
  flour

Beat egg yolks until light. Add milk, oil, and salt. Add combined flours and mix until smooth. In another bowl, beat egg whites until stiff. Fold into batter. Drop by spoonfuls into hot, oiled iron muffin or gem pans. Bake in 325° oven 20 to 25 minutes, until golden brown. Makes 18.

Soy Cereals

Soybeans are a welcome addition to any meal, but they are especially useful at breakfast, where protein needs often are neglected.

It's a simple matter to add soy grits or soy flour to the old breakfast standbys, such as fried mush. But do try something new, such as making your own breakfast cereal. Not only is it an easy, fix-in-a-hurry breakfast full of protein, vitamins, and minerals, but homemade soy cereals are the best tasting cereals you've ever eaten and for less than half the price of the commercial ones.

SOY-WHEAT CEREAL

¼ cup vinegar
1¼ cups soy milk (page 33)
1 cup dark molasses
¼ teaspoon salt
1 teaspoon vanilla

½ teaspoon maple flavoring
1 teaspoon soda
¼ cup boiling water
2½ cups whole wheat flour
½ cup soy flour (page 29)

Add vinegar to soy milk. Let set 10 minutes to curdle. Add molasses, salt, vanilla, maple flavoring, and soda which has been dissolved in boiling water. Blend in flours. Pour into greased, shallow cake or cookie pans. Bake 40 minutes in 375° oven. Remove from pans and wrap in slightly damp cloth until cold. Grind in meat grinder, using coarse blade. Spread on ungreased cookie sheets and toast in 300° oven until golden brown, about 15 to 20 minutes. Serve as a breakfast cereal, with sugar and soy milk.

FRIED SOY MUSH

1 cup cornmeal
1 cup cold water
2½ cups boiling water
1 teaspoon salt

2 tablespoons soy grits (page 27)
4 tablespoons wheat flour
4 tablespoons cooking or salad oil

Mix cornmeal and cold water. Add boiling water, salt, and soy grits. Cook in a double boiler over hot water one hour. Pour into an 8 × 5-inch loaf pan which has been rinsed with cold water. Chill until set. Run knife along the edge to loosen and unmold on a clean cutting board. Cut into one-half inch slices and dredge each slice in flour. Brown in hot cooking oil. Makes eight servings.

SPROUT CEREAL

1 cup dried sprouts (page 41)
1 cup sunflower seeds
Juice of 3 lemons
3 tablespoons nutritional yeast
3 tablespoons honey

½ cup soy flour (page 29)
6 cored and seeded apples
  (unpeeled) grated
½ cup roasted soy grits (page 48)

Combine all ingredients. Serves six.

CARROT CEREAL

½ cup brown sugar
1 egg
1 cup cooked carrots, mashed
  and cooled
1½ cups wheat flour

½ cup soy flour (page 29)
½ teaspoon salt
1½ teaspoons baking powder
¾ cup seedless raisins
1 teaspoon vanilla

Combine all ingredients. Beat well and spread thinly on a greased shallow baking pan. Bake in 300° oven until lightly browned. Cool and crumble with the hands, then spread on cookie sheets. Return to 250° oven, stirring occasionally, until well browned and crisp. Store in tightly covered container.

SOY-OAT CEREAL

1 cup soy flour (page 29)
1 cup all-purpose white wheat
    flour
2 cups whole wheat flour
2 cups rolled oats

1 cup cornmeal
1½ cups soy milk (page 33)
1½ cups light cream
1½ teaspoons salt
¼ cup sugar

Mix ingredients and knead well. Roll out very thin on a lightly floured board. Place thin dough on ungreased cookie sheets and cut into squares. Bake in a 250° oven 45 minutes or until golden brown and crisp. Break up with the hands or grind through the coarse blade of a food grinder. Makes two quarts of cereal.

CRUNCHY OAT CEREAL

½ cup brown sugar
1 egg
1 cup wheat flour
½ cup soy flour (page 29)
1¼ cups rolled oats

¼ teaspoon salt
2 teaspoons baking powder
1 cup coconut, grated
1 teaspoon vanilla

Combine all ingredients and beat well. If dough is stiff, add 1 table-spoon water. Spread thinly on greased cookie sheet. Bake in 300° oven until lightly browned. Cool, then crumble with the hands. Spread again on cookie sheet and toast in 250° oven, stirring occasionally, until well browned and crisp. Store in tightly covered container.

GRANOLA

3 cups uncooked rolled oats
1 cup roasted soy grits (page 48)
½ cup sunflower seeds
1 cup coconut, shredded
1 teaspoon salt

½ cup honey
¼ cup cooking or salad oil
1 teaspoon vanilla
½ cup seedless raisins
½ cup dates, chopped

Combine oats, soy grits, sunflower seeds, coconut, and salt. Drizzle with a mixture of the honey, oil, and vanilla. Mix well and spread on a greased cookie sheet. Toast in a 250° oven 1 to 1½ hours, stirring occasionally, until light brown and crisp. Cool, then add raisins and dates. Mix well. Makes about seven cups of cereal. Eat as a snack or with soy milk as a breakfast cereal.

WHOLE GRAIN CEREAL

(To be cooked as a hot cereal)

1 pound soy grits (page 27)
1 pound rolled oats

1 pound wheat germ

Combine grains and store in the refrigerator in a tightly sealed jar. To cook, add one cup of mixture to 3 cups boiling water in which one-half teaspoon salt has been dissolved. Cover, lower heat and cook 20 to 30 minutes. Serves six.

Soy Desserts

Desserts made from soybeans and soybean products are as varied as the main dish uses of soybeans.

Use soy flour in cakes and pies and give a nutritional boost to your puddings and cakes with soy milk. As for versatility, you can't beat soy curd. It can be used in everything from cheesecake to dessert toppings.

SOY STRUDEL

Dough:

2½ cups all-purpose white
    wheat flour
½ cup soy flour (page 29)
½ teaspoon salt

1 tablespoon cooking or salad
    oil
1 egg, well beaten
1 cup lukewarm water

Sift flours and salt into a bowl. Make a well in the center and put oil and egg in the depression. Work flour gently into the oil and egg. Add water to make a soft, sticky dough. Turn onto lightly-floured board, throwing dough at the board about 100 times, or until dough is smooth and no longer sticky. Knead lightly and form into a ball. Lightly brush with cooking oil. Rinse a heavy bowl in hot water, invert over dough on kneading board and let set 30 minutes.

To stretch dough, spread a clean cloth over a table large enough to make a 3 × 5-foot working surface. Sprinkle the cloth with ½ cup flour. On this surface, roll out dough to a large rectangle, rolling out as thin as possible and lifting dough occasionally to keep from sticking. Brush lightly with cooking oil. Being careful not to tear it, stretch dough until it is paper thin and hangs over the edge of the table on all sides. With kitchen scissors, trim off the thicker edges. Allow to dry 10 minutes.

Filling:

½ cup butter or margarine,
   melted and cooled
½ cup soy nuts (page 47),
   finely ground
2 egg yolks
¼ cup sugar

½ teaspoon salt
1 pound fresh soy curd
   (page 35)
¼ cup seedless raisins
½ teaspoon vanilla
½ teaspoon grated lemon rind

Brush the butter or margarine over the entire suface of the dough. Sprinkle with the ground soy nuts. In a blender or electric mixer, beat egg yolks, sugar, and salt until thick. Gradually add soy curd, blending after each addition. Mix until smooth. With a spoon, stir in raisins, vanilla and lemon rind. Spoon mixture carefully over dough and spread with spatula. Fold over edges on three sides of dough on table. Roll up loosely. Cut roll in half and gently slide each piece onto a greased baking sheet. Brush surface with melted butter or margarine and bend into a horseshoe shape. Bake in a 350° oven 35 to 45 minutes, until golden brown. Baste with melted butter or margarine several times during the baking. Cool slightly and dust with confectioner's sugar. Cut into two-inch slices. Makes 12 to 15 servings.

APPLESAUCE CAKE

1 cup soy flour (page 29)
1½ cups all-purpose white
   wheat flour
½ teaspoon salt
½ teaspoon powdered cinnamon
½ teaspoon powdered cloves
½ teaspoon powdered allspice
1 cup seedless raisins

½ cup soy nuts (page 47)
½ cup dates, chopped
½ cup solid vegetable shortening
1 cup sugar
2 eggs
2 cups unsweetened applesauce
2 teaspoons soda
½ cup boiling water

In a medium-sized bowl, combine soy flour, white flour, salt, cinnamon, cloves, allspice, raisins, soy nuts, and dates. Mix well and set aside. In a large mixing bowl, cream shortening and sugar until fluffy.

Add eggs and beat well by hand or with electric mixer. Blend in apple-sauce. In a measuring cup, dissolve soda in boiling water and add to applesauce mixture alternately with the combined dry ingredients and fruit, mixing in a little of each at a time, and beating well after each addition. Pour the batter into a greased 8 × 12-inch baking pan and bake in 350° oven one hour, until lightly browned and a toothpick inserted comes out clean. Serve warm without icing or frost with vanilla or caramel icing. Makes 12 servings.

BLINTZES

Batter:

1 cup all-purpose white wheat flour

1 teaspoon salt

4 eggs, well beaten

1 cup soy milk (page 33)

Combine flour and salt. Mix together eggs and soy milk. Add milk mixture to dry ingredients and mix until smooth. Lightly grease a six-inch skillet. Heat until a drop of water "dances" across the skillet. Pour out enough batter to make a very thin cake, tilting the skillet to spread the batter evenly. Cook over low heat on one side only until top of cake is dry and blistered. Turn out on cloth and allow to cool, cooked side up. Repeat until all batter is used.

Filling:

1½ pounds soy curd (page 35)

2 egg yolks, well beaten

1 tablespoon butter, melted

Salt, sugar, and cinnamon to taste

In a blender, combine soy curd, egg yolks, and butter until smooth. Season to taste with salt, sugar, and cinnamon. Place a tablespoon of mixture in the center of each blintz on the browned side. Fold edges over to form an envelope.

Blintzes may be prepared and filled in advance and kept in the refrigerator until ready to serve. Just before serving, fry in butter or oil until brown on all sides. Or bake until golden in a 350° oven. Serve hot with sour cream or sprinkled with a mixture of ½ cup sugar and 1 teaspoon cinnamon. Makes 10 blintzes.

SOY PEACH SHORTCAKE

1½ cups all-purpose white
    wheat flour
¾ cup soy flour (page 29)
3 teaspoons baking powder
½ teaspoon salt
¼ cup brown sugar, packed
½ cup solid vegetable shortening
½ cup pecans, chopped

1 egg, well beaten
⅔ cup soy milk (page 33)
2 cups fresh peaches, peeled,
    sliced and sweetened
1 cup fresh soy curd (page 35)
¼ cup confectioner's sugar
½ teaspoon vanilla

Combine both flours, baking powder, salt, and brown sugar in a bowl. Using a pastry blender or two knives, cut in shortening until the mixture resembles coarse meal. Stir in pecans. In another bowl, combine egg and soy milk and add, all at once. Mix only until blended. Turn out on a floured board and knead a few times. Roll to one-half inch thickness and cut into three-inch squares. Place on a greased baking sheet and bake 10 to 12 minutes in a 425° oven. Split shortcakes. While still warm, layer peaches and shortcakes and top with the soy curd which has been flavored with confectioner's sugar and vanilla. Makes six servings.

CURD CHEESECAKE

1¼ cup graham cracker crumbs
¼ cup soy flour (page 29)
½ cup butter or margarine,
    melted
3 tablespoons confectioner's
    sugar
1 cup crushed pineapple,
    well drained
3 eggs, separated
1 cup granulated sugar

Grated rind and juice of 1 lemon
½ teaspoon salt
½ teaspoon grated nutmeg
¼ cup all-purpose white
    wheat flour
¼ cup soy flour (page 29)
1 cup heavy cream, whipped
1 teaspoon vanilla
1 pound (about 2 cups) fresh
    soy curd (page 35)

Combine graham cracker crumbs with ¼ cup soy flour, butter or margarine, and confectioner's sugar. Press one half of this mixture on the bottom and sides of a well-greased round cake pan. Spread well-drained pineapple over crumbs. In a mixing bowl, beat egg yolks until light, then gradually add granulated sugar, beating thoroughly. Add grated rind and juice of lemon, salt, nutmeg, white flour, ¼ cup soy flour, and whipped cream to which vanilla has been added. Fold in soy curd and beat well to a smooth batter. Pour over pineapple and spread remaining crumbs over the top. Bake in a 275° oven one hour, until firm. Turn off heat, open oven door and let stand in oven one hour, until cool. Invert carefully onto serving plate. Serves eight.

SOY – WHEAT GERM CAKE

¼ cup solid vegetable shortening
½ cup sugar
½ cup molasses
1 egg, well beaten
¾ cup soy milk (page 33)
2 tablespoons orange juice
1 teaspoon grated orange rind
1¼ cups all-purpose white
    wheat flour

¾ cup soy flour (page 29)
¼ teaspoon soda
1 teaspoon baking powder
¼ teaspoon salt
¼ teaspoon powdered cinnamon
1/8 teaspoon powdered ginger
½ cup wheat germ

Cream shortening and sugar until fluffy. Add molasses and egg. Beat well. One at a time, add milk, orange juice, and rind, beating well after each addition. In another bowl, combine remaining ingredients. Add, all at once, to batter, blending in well. Do not overbeat. Pour into greased 8-inch square cake pan and bake in 350° oven 30 minutes, or until top springs back when touched lightly in center. Cool and frost with icing without removing from cake pan. Serve from pan. Serves eight.

SOY NUT CRUMB
PIE SHELL

½ cup soy nuts (page 47),
    finely ground

½ cup graham cracker crumbs
⅓ cup cooking or salad oil

Combine all ingredients. Press onto sides and bottom of a 9-inch pie pan. Chill before filling.

SOY CURD — YOGURT CAKE

Soy nut crumb pie shell
  (next recipe)
1 cup soy yogurt (page 00)
3 eggs, separated
1 teaspoon vanilla
1 tablespoon lemon juice

1 tablespoon grated lemon rind
⅓ cup honey
¼ teaspoon salt
¼ cup whole wheat flour
1 pound (about 2 cups) fresh
  soy curd (page 35)

Pat crumb shell into bottom of a 9-inch round cake pan. In a blender, combine yogurt, egg yolks, vanilla, lemon juice, lemon rind, honey, salt, flour, and curd. Blend until smooth. Beat egg whites with a rotary beater until they form stiff peaks. Fold in blender mixture and turn into crumb pie shell. Bake in 350° oven until center is firm. Cool completely before removing from pan. Serves eight.

SOY PIE CRUST

½ cup roasted soy flour
  (page 49)
1½ cups wheat flour

½ teaspoon salt
⅔ cup solid vegetable shortening
Cold soy milk (page 33)

Combine soy flour, wheat flour, and salt. Blend in shortening, using a pastry blender or two knives, until mixture is the consistency of coarse meal. Add cold soy milk, one tablespoon at a time, until a firm dough is formed. Yield: One double crust pie or two 9-inch pie shells.

CHOCOLATE PIE CRUST

1 cup wheat flour
½ cup soy flour (page 29)
2½ tablespoons cocoa
2 teaspoons sugar

½ teaspoon salt
½ cup solid vegetable shortening
Cold soy milk (page 33)

Combine wheat flour, soy flour, cocoa, sugar, and salt. Mix well. Cut in shortening with a pastry blender or two knives until mixture is the consistency of coarse meal. One tablespoon at a time, add just enough cold soy milk to form a stiff dough. Pat into a ball, chill, and roll out to fit a 9-inch pie pan. Fold back edges to form a fluted rim and pierce bottom and sides with a fork. Bake 12 minutes in a 375° oven. Makes one 9-inch crust.

BANANA CREAM PIE

3½ tablespoons cornstarch
¼ cup soy flour (page 29)
½ cup sugar
¼ teaspoon salt
2 cups soy milk (page 33)

3 eggs, separated
1 teaspoon vanilla
2 bananas
Baked soy pie crust (page 80)
2 tablespoons sugar

Combine cornstarch, soy flour, ¼ cup sugar, and salt. Gradually add soy milk, stirring to mix well and dissolve the dry ingredients. Cook in the top of a double boiler, stirring occasionally, until thickened. In a small bowl, add remaining sugar to the 3 egg yolks. Stir in a small amount of the hot milk mixture and mix well. Add eggs to cooked mixture. Stir well and cook five minutes longer, stirring frequently. Remove from heat and add vanilla. Set aside to cool slightly. Meanwhile, slice one banana into the bottom of the baked pie crust. Pour hot mixture over banana slices and top with remaining banana, sliced. In a small bowl, beat the 3 egg whites until they form soft peaks. Gradually add remaining sugar, beating until stiff. Spread over top of banana-cream mixture. Bake in 375° oven 10 minutes, or until golden brown. Serves six to eight.

CHOCOLATE PIE

3 squares (3 ounces)
  unsweetened chocolate
2½ cups soy milk (page 33)
3 tablespoons wheat flour
3 tablespoons soy flour (page 29)
3 tablespoons cornstarch
¾ cup sugar

¼ teaspoon salt
3 eggs, separated
1 tablespoon butter or margarine
2 teaspoons vanilla
2 tablespoons sugar
1 soy pie crust, baked

Melt chocolate in soy milk in top of double boiler. Blend with egg beater. In a bowl, combine wheat flour, soy flour, cornstarch, sugar, and salt. Stir gradually into hot milk mixture and cook 15 minutes, stirring constantly until thickened. Stir a small amount of the hot mixture into the well-beaten egg yolks and blend well, then return to the hot mixture. Cook five minutes longer. Add butter and vanilla and set aside to cool. Meanwhile, make meringue by beating egg whites until they form soft peaks. Gradually add sugar, beating well until stiff. Pour slightly cooled chocolate mixture into baked pie crust and spread meringue in swirls over the top. Bake in a 375° oven 10 minutes, or until golden brown. Serves six to eight.

FRUIT-CURD PIE

1 soy pie crust, unbaked
1½ cups raw apples, peeled
  and thinly sliced
3 tablespoons sugar
½ teaspoon powdered cinnamon
¼ teaspoon grated nutmeg
2 eggs, slightly beaten

¾ cup pressed soy curd
  (page 37), chopped
½ cup sugar
½ cup fresh soy curd (page 35)
1/8 teaspoon salt
1 teaspoon grated lemon rind

Line pie crust with apple slices. Combine 3 tablespoons sugar, cinnamon, and nutmeg, and sprinkle over apples. In a bowl, combine eggs, chopped soy curd, ½ cup sugar, fresh soy curd, salt, and rind. Pour over apples and bake in a 425° oven 10 minutes. Reduce heat to 350° and bake 30 minutes longer. Serves six to eight.

LEMON CURD PIE

Double soy pie crust recipe,
  unbaked
3 cups fresh soy curd (page 35)
¼ cup wheat flour
¼ cup soy flour (page 29)
2 tablespoons grated orange rind
2 tablespoons grated lemon rind

1 tablespoon vanilla
1/8 teaspoon salt
4 eggs
1 cup sugar
2 tablespoons sifted
  confectioner's sugar

Line 9-inch pie pan with one-half of pie crust dough. Roll out remaining dough and cut into 1-inch strips. Combine soy curd, wheat flour, soy flour, orange and lemon rind, vanilla, and salt. Beat eggs until foamy. Add sugar gradually, beating until thick. Stir into curd mixture until smooth and well blended. Pour into unbaked pie crust and cover with woven pastry strips. Bake in 325° oven 50 to 60 minutes, until filling is firm and crust is golden brown. Cool. Sprinkle with confectioner's sugar. Serves six to eight.

PINEAPPLE-CURD PIE

1½ cups fresh soy curd (page 35)
2 eggs, separated
¼ cup butter or margarine,
  melted
½ cup sugar
¼ teaspoon salt

1 tablespoon all-purpose white
  wheat flour
Grated rind of 1 lemon
¼ cup soy milk (page 33)
1 soy pie crust, unbaked

Combine soy curd, egg yolks, butter or margarine, sugar, salt, wheat flour, lemon rind, and soy milk. Blend well. Beat egg whites until stiff and fold in. Pour into unbaked pie crust and bake 10 minutes in 425° oven. Reduce heat to 325° and bake 40 minutes longer, until filling is firm. Spread with Pineapple Glaze (page 84) and return to oven five minutes. Serves six to eight.

Pineapple Glaze:

¾ cup pineapple juice (drained from canned pineapple)
½ cup sugar

2 tablespoons cornstarch
1 (No. 2) can crushed pineapple, drained

Combine pineapple juice, sugar, and cornstarch. Cook over low heat, stirring constantly, until thick and clear. Add pineapple and cool. Spread over curd filling.

PUMPKIN PIE

¼ cup soy flour (page 29)
1¼ cups soy milk (page 33)
1½ cups cooked pumpkin
½ cup brown sugar
½ teaspoon salt
1 teaspoon powdered cinnamon

¼ teaspoon powdered allspice
½ teaspoon grated nutmeg
¼ teaspoon powdered cloves
2 eggs, slightly beaten
1 soy pie crust, baked

Combine soy flour with ¼ cup soy milk. Mix with pumpkin and remaining soy milk, brown sugar, salt, and spices. Heat over hot water. Add beaten eggs. Mix well. Pour hot filling into pre-baked soy crust and bake in 350° oven 30 minutes, or until filling sets. Serves six to eight.

SOY SPROUT PIE

¾ cup dark brown sugar
1 large (13-ounce) can evaporated milk
½ teaspoon salt
1 teaspoon powdered cinnamon

½ teaspoon grated nutmeg
1½ cups dried soy sprouts (page 41)
2 eggs, beaten slightly
1 soy pie crust, unbaked

Combine brown sugar, one third of the milk, salt, and spices. Add sprouts, eggs, and remainder of the milk. Pour into pie crust and bake in 350° oven 45 minutes, or until knife inserted in center comes out clean.

STRAWBERRY CURD PIE

1½ cups fresh soy curd (page 35)   1 quart fresh strawberries
1½ cups sugar   1½ cups water
1 teaspoon vanilla   Red food coloring
1 soy pie crust, baked   3 tablespoons cornstarch

Blend soy curd with ½ cup sugar and ½ teaspoon vanilla. Spread
one-third of the mixture over the bottom of baked pie crust. Fill the
shell with about one half of the best of the whole strawberries. Cook
the remaining berries in water until soft. Force through a strainer and
color with food coloring. In a saucepan, combine remaining 1 cup
sugar and cornstarch. Add to juice. Cook, stirring constantly, until
thickened. Pour over berries in pie crust. Refrigerate to chill thoroughly.
Just before serving, top with remaining two-thirds of soy curd mixture.
Serves six to eight.

SOY CUSTARD

2 cups soy milk (page 33)   2 eggs, well beaten
¼ cup sugar   ½ teaspoon vanilla
¼ teaspoon salt

Combine milk, sugar, salt, and beaten eggs. Cook in double boiler
until mixture coats spoon. Cool slightly and add vanilla. Chill. Serves
four.

PINEAPPLE-COCONUT SOY CURD

1 cup fresh soy curd (page 35),   1 cup fresh or packaged grated
  beaten until fluffy   coconut
1 cup canned or fresh pineapple,   4 tablespoons honey
  diced and drained   1 teaspoon vanilla

Combine all ingredients. Chill. Serves six.

SOY CARROT PUDDING

2 cups cooked carrots, mashed
1 cup soy pulp (page 29)
1 cup pitted dates, chopped
1 cup seedless raisins
½ teaspoon salt

2 teaspoons powdered cinnamon
2 eggs, well beaten
1 cup apple juice
2 cups soy nuts (pages 47),
    finely ground

Combine all ingredients in order, mixing well. Bake in two-quart oiled casserole 30 minutes in 350° oven or 20 minutes in oiled custard cups. Serves six to eight.

RICE PUDDING

½ cup rice, uncooked
4 cups soy milk (page 33)
4 tablespoons honey
2 tablespoons butter or margarine

¼ cup seedless raisins
½ teaspoon salt
½ teaspoon powdered cinnamon
1 teaspoon vanilla

Combine ingredients and cook, covered, over very low heat, in an electric slow cooker or in a 300° oven until soy milk is absorbed and rice is soft. Serves six.

APPLE-BREAD PUDDING

2 cups soft bread crumbs
½ cup soy flour (page 29)
½ cup brown sugar
½ teaspoon grated nutmeg
½ teaspoon powdered cinnamon
¼ teaspoon salt

6 cups tart apples, peeled and
    sliced
2 tablespoons lemon juice
1 teaspoon grated lemon rind
½ cup soy milk (page 33)
1 cup soy curd topping (page 88)

Combine bread crumbs, soy flour, brown sugar, spices, and salt. Spread one-third of the mixture in the bottom of a greased 1½-quart casserole or baking pan. Spread one-half the apples over the crumb mixture. Sprinkle with one-half of the lemon juice and grated rind.

Repeat layers and top with crumb mixture. Pour soy milk over top and bake 30 minutes in 375° oven. Serve warm topped with soy curd topping. Serves four to six.

SOY ICE CREAM

2 cups soy milk (page 33)
¾ cup honey
¼ teaspoon salt
2 eggs, well beaten

2 tablespoons soy flour (page 29)
1 cup heavy cream, whipped
2 teaspoons vanilla

Heat soy milk almost to boiling. Add honey and salt and cool. Gradually add eggs and soy flour, beating well. Cook over low heat, stirring constantly, until mixture thickens. Cool. Fold in whipped cream and vanilla. Freeze until firm in ice cream freezer or freezing section of refrigerator. Serves six.

SOY SHERBET

1½ cups fresh soy curd (page 35)
1 tablespoon lemon juice
1 teaspoon grated lemon rind
4 tablespoons honey

1½ cups soy yogurt (page 54)
1/8 teaspoon grated nutmeg
1/8 teaspoon powdered cinnamon

Blend all ingredients until smooth in blender. Pour into ice cube trays or a shallow pan and freeze in freezing section of refrigerator. Stir occasionally as it freezes. Serves six.

FROZEN SOY YOGURT

2 cups soy yogurt
   (page 54)

2 cups fresh or frozen fruit
   (berries or peaches), mashed
   and drained

Spread yogurt in ice cube tray. Freeze to the mushy stage. Beat well, then blend in mashed fruit. Freeze again to mushy stage, then beat again. Freeze until firm. Serves six.

SOY APPLE CRISP

4 cups apples, peeled and
  sliced
¾ cup granulated sugar
¼ teaspoon powdered cinnamon

½ cup brown sugar
⅔ cup soy flour (page 29)
¼ cup butter or margarine

Combine apples, granulated sugar, and cinnamon. Place in a greased, shallow 8-inch square pan or 1½-quart casserole. In a bowl, mix brown sugar and soy flour. Using a pastry blender or two knives, work in butter or margarine until crumbly. Sprinkle over apple mixture. Bake uncovered 25 minutes in 375° oven. Serves four to six.

SOY CURD TOPPING

Beat fresh soy curd (page 35) until it is fluffy. Sweeten to taste with honey or sugar and flavor with vanilla, lemon juice or cinnamon. Chill and serve as a topping on baked apples, fresh fruit, or shortcake. May be used as a quick icing for cake.

SOY NUT TOPPING

4 tablespoons sugar
1½ teaspoons powdered
  cinnamon
4 tablespoons soy grits (page 27)

1 cup fine dry bread crumbs,
  sifted
4 tablespoons butter or
  margarine, melted

Combine all ingredients. Spread over cookie sheet. Toast in a 250° oven one hour, or until crisp and golden brown. Use to top puddings, fruit pies or ice cream sundaes.

Soybeans with Eggs

You'll have to beat a lot of bushes before you'll scare up a good crowd of egg-boosters. They may be blessed with the best of proteins, but eggs need all the help they can get to give oomph to their flavor and texture. Soybeans make good egg partners with their mild, nutty flavor and firm texture. They step up the protein level and add a dash of interest. You can introduce them in many ways—cooked, mashed, ground, roasted, sprouted or in curd form.

EGGS FOO YOONG

½ cup mushrooms, sliced
(canned or fresh)
1 tablespoon butter or margarine
1 cup chicken broth
2 tablespoons soy sauce
¼ cup water
1 tablespoon cornstarch

2 tablespoons cooking or salad
oil
6 eggs
1 cup cooked chicken,
chopped fine
⅔ cup onion, thinly sliced
2½ cups soy sprouts (page 38)

Sauté mushrooms in butter or margarine. Set aside. Heat broth to simmering in a small saucepan. Add soy sauce. In a cup, blend water into cornstarch. Pour mixture slowly into simmering broth, stirring constantly until thickened. Keep warm over hot water. Meanwhile, heat oil in heavy skillet over moderate heat. Beat eggs with beater until soft peaks form. Stir in chicken, onions, sprouts, and mushrooms. Pour one-half cup of egg mixture into heated oil in skillet. Over medium heat, fry until lightly browned on one side. Flip and brown on other side. Repeat until all of mixture is fried. Serve with sauce. Makes 12 patties.

SOY SCRAMBLED EGGS

6 eggs
⅓ cup soy milk (page 33)
½ cup soy pulp (page 29)

½ teaspoon salt
Pepper
2 tablespoons butter or margarine

Beat eggs until frothy. Stir in soy milk and soy pulp. Mix well. Season with salt and pepper. Heat butter or margarine in heavy skillet over medium heat. Pour in egg mixture, all at once, and cook slowly, stirring gently, until eggs are set but not dry. Makes six servings.

SOY GRIT OMELET

4 eggs, separated
¼ cup soy grits (page 27)
½ cup water

½ teaspoon salt
1 teaspoon solid vegetable
  shortening

Beat egg yolks until thick. Combine soy grits with ¼ cup water and add to yolks. Add salt and the remaining ¼ cup water to egg whites. Beat until stiff peaks form. Gradually fold egg yolk mixture into the stiffly beaten egg whites, then pour into a skillet in which the shortening has been heated. Cover and cook over low heat until set around the edges. Remove lid and put in a 300° oven to slightly brown top. Makes four to six servings.

EGG-SPROUT PATTIES

2 small onions, finely chopped
2 green peppers, finely chopped
2 cups soy sprouts (page 38)

½ teaspoon salt
4 eggs
3 tablespoons cooking or salad oil

Combine first five ingredients. Beat well. Drop by tablespoonfuls onto skillet in which oil has been heated. Brown patties on both sides. Serves four.

SPROUT SOUFFLÉ

6 slices bread
1 cup roasted soy sprouts
  (page 49), chopped
¾ teaspoon salt

¼ teaspoon dry mustard
¼ teaspoon paprika
2 cups soy milk (page 33)
5 eggs, well beaten

Arrange bread in layers in a greased casserole, sprinkling chopped soy sprouts over each layer. Reserve some sprouts for the top. Add seasonings and soy milk to eggs and pour over bread layers. Sprinkle reserved sprouts over top. Bake in 350° oven about one hour, until souffle is puffy and a knife inserted in center comes out clean. Serves six.

Soy Salads

With a little imagination, soybeans in some form can perform many magic tricks with your main or side dish salads. Cooked green or dried soybeans may be used in bean or vegetable salads. Soy sprouts may be added to tossed, layered, or molded salads. Ground soy nuts make delicious toppings for fruit salads.

Soy curd may be substituted in almost any salad which lists cottage cheese as an ingredient. Pressed curd may be chopped and added to fruit salads and fresh curd may be thinned, flavored, and used as a salad dressing. The soyloaf meat substitutes we've described may be combined with celery, pickle, and mayonnaise for a main-dish, meat-like salad.

SOY BEAN SALAD

2 cups cooked fresh soybeans,
  chilled (page 22)
¼ cup sweet pickles, finely
  chopped
2 tablespoons onion, finely
  chopped

¼ cup celery, finely chopped
2 hard-cooked eggs, chopped
1/8 teaspoon salt
Mayonnaise or salad dressing
Lettuce leaves

Combine first six ingredients. Add mayonnaise to taste. Toss lightly.
Serve on lettuce leaves.

SOYBEAN POTATO SALAD

3 cups cooked fresh soybeans
  (page 22)
1 cup celery, diced
4 hard-cooked eggs, diced
4 green onions, sliced
¼ cup sweet pickle relish
¼ cup pimiento, diced

1 cup cold cooked potato, diced
½ cup mayonnaise or salad
  dressing
1 tablespoon lemon juice
1 teaspoon prepared horseradish
Salt and pepper
Paprika

Combine first seven ingredients. Blend mayonnaise, lemon juice, and
horseradish. Season to taste with salt and pepper. Mix with salad
ingredients to coat well. Add more mayonnaise if desired. Garnish
with paprika. Serves six to eight.

SOY FRUIT SALAD

½ cup seedless raisins
1 cup apple, peeled and chopped
1 cup cooked fresh soybeans
  (page 22)

½ cup celery, chopped
1 medium avocado, diced
Mayonnaise or salad dressing
Lettuce

Combine first five ingredients. Moisten with mayonnaise or salad
dressing and serve on lettuce. Serves six.

HOT BEAN SALAD

4 slices bacon, fried and
  crumbled
1 small onion, thinly sliced
2 cups cooked, dried soybeans,
  drained (page 24)
2 cups cooked fresh soybeans,
  drained (page 22)

2 cups cooked or canned green
  beans, drained
½ cup brown sugar
½ teaspoon salt
¼ cup vinegar

Combine all ingredients, folding to mix well. Put into a covered casserole dish and bake 30 minutes in a 350° oven. Uncover and bake 10 minutes longer. Serves six to eight.

SOYBEAN FRUIT BALLS

1½ cups soy pulp (page 29)
½ teaspoon salt
6 teaspoons lemon juice
3 tart apples, peeled and finely
  diced

¼ cup currants or seedless
  raisins
½ cup soy nuts (page 47),
  coarsely ground
Lettuce
Mayonnaise or salad dressing

To the soy pulp add salt and lemon juice. Mix well. Add diced apples and currants or raisins and shape into balls. Roll in ground soy nuts. Serve on lettuce with mayonnaise or salad dressing.

FRESH SOYBEAN SALAD

1½ cups cooked fresh soybeans
  (page 22), drained
½ cup celery, thinly sliced
1 tablespoon onion, minced

2 tablespoons pimiento, chopped
1 hard-cooked egg, chopped
Mayonnaise or salad dressing

Combine vegetables and chopped egg. Add mayonnaise or salad dressing to moisten. Serves four.

CHICKEN-SOY SALAD

2 cups cooked chicken, chopped
½ cup celery, diced
½ cup pitted ripe olives,
   sliced

½ cup soy nuts (page 47),
   coarsely ground
Mayonnaise or salad dressing
Lettuce

Combine chicken, celery, olives and soy nuts. Moisten with mayonnaise or salad dressing and serve on lettuce leaves or in tomato shells. Serves four to six.

CURD-PINEAPPLE SALAD

1 envelope (1 tablespoon)
   unflavored gelatin
¼ cup cold water
½ cup sugar

2 cups boiling water
1 cup canned crushed pineapple,
   drained
1 cup pressed soy curd (page 37),
   diced

Let gelatin set in cold water to soften. Dissolve sugar and softened gelatin in boiling water. Add pineapple and curd. Stir well and pour into a mold which has been rinsed with cold water. Chill until set. Serves six.

STUFFED AVOCADO SALAD

2 medium avocados
1 tablespoon lemon juice
1 cup fresh soy curd (page 35)
1 teaspoon onion, minced

½ teaspoon celery salt
Lettuce
Mayonnaise or salad dressing

Cut avocados in half lengthwise and peel. Hollow out stem ends slightly. Brush inside with lemon juice. Blend soy curd, onion, and celery salt, and beat until smooth. Fill avocado halves with mixture and press halves together. Wrap in waxed paper and chill thoroughly. Cut into slices lengthwise and serve slices on lettuce leaves. Top with mayonnaise or salad dressing. Makes six salads.

THREE-BEAN
SOYBEAN SALAD

2 cups cooked fresh soybeans
  (page 22), drained
2 cups canned kidney beans,
  with liquid
2 cups green beans, drained
1 medium onion, thinly sliced

½ green pepper, chopped
2 cloves garlic
1 teaspoon celery seed
½ cup vinegar
½ cup salad oil
½ cup sugar

Combine soybeans, kidney beans with liquid, green beans, onion slices, and green pepper in a large bowl. Add garlic and sprinkle all with celery seed. In a small saucepan combine vinegar, oil and sugar. Heat to dissolve sugar, then pour over bean mixture. Blend well. Refrigerate eight hours or more. Remove garlic before serving. Serves eight.

STUFFED CELERY

1 cup fresh soy curd (page 35)
2 tablespoons French dressing
2 tablespoons stuffed green
  olives, chopped

12 ribs celery, cut in serving-
  size pieces

Blend soy curd and French dressing until smooth. Fold in chopped olives and use to fill celery pieces.

WALDORF SALAD
WITH SOY NUTS

2 cups unpeeled apple, chopped
2 cups celery, chopped
1 cup soy nuts, (page 47),
  chopped

Mayonnaise or salad dressing
Lettuce

Combine apple, celery and soy nuts. Moisten with mayonnaise or salad dressing and serve on lettuce. Serves six.

FROZEN CURD SALAD

1½ cups fresh soy curd (page 35)
1 tablespoon mayonnaise or
  salad dressing
1 tablespoon lemon juice
¼ teaspoon salt
2 oranges, diced
2 bananas, diced
1 cup canned crushed pineapple,
  drained

1 red maraschino cherry, sliced
2 green maraschino cherries,
  sliced
1 cup heavy cream, whipped
¼ cup sugar
½ teaspoon vanilla
Salad greens

Combine soy curd, mayonnaise, lemon juice, and salt in blender until smooth and creamy. Stir in fruits and fold in whipped cream to which sugar and vanilla have been added. Freeze. Slice and serve on salad greens. Makes six servings.

SALMON CURD SALAD

1 cup canned salmon, drained
  and flaked
1 cup cooked soy curd (page 35),
  chopped

½ cup celery, chopped
¼ cup sweet pickle relish
Mayonnaise or salad dressing

Combine first four ingredients. Add enough mayonnaise to moisten. Serves four.

LUNCHEON SALAD

1 cup unpeeled apple, diced
1 cup pineapple, diced
1 cup fresh peaches, diced

1 cup fresh soy curd (page 35)
¼ cup sour cream
Lettuce leaves

Combine fruits, curd, and sour cream. Serve on lettuce leaves. Serves six.

CURD-VEGETABLE SALAD

1 clove garlic
2 cups fresh soy curd
  (page 35)
1 teaspoon salt
¼ teaspoon paprika
2 teaspoons lemon juice
2 tablespoons green onion,
  chopped

2 tablespoons pimiento, chopped
¼ cup celery, chopped
Salad greens
2 cucumbers, sliced
1 medium onion, sliced
2 large tomatoes, quartered
2 carrots, sliced
French dressing

Rub salad bowl with cut clove of garlic. Add soy curd, salt, paprika, lemon juice, green onion, pimiento, and celery. Toss, cover and chill. To serve, unmold in center of large salad plate; surround with salad greens topped with arrangement of sliced vegetables. Serve with French dressing. Serves six to eight.

SEAFOOD-SPROUT SALAD

2 cups soy sprouts (page 38),
  cooked, chilled and well-
  drained
2 cups green onions, finely
  chopped (tops and all)
1 cup cooked shrimp or crab
  meat, cut up

1 cup soy yogurt (page 54)
1 teaspoon curry powder
1 clove garlic, crushed
1 tablespoon lemon juice
2 tablespoons soy sauce
1 teaspoon sugar

Combine soy sprouts with onion and shrimp or crab meat. Set aside to chill. Meanwhile, make dressing by mixing yogurt, curry powder, garlic, lemon juice, soy sauce, and sugar. Beat until smooth, then remove garlic and discard. Pour dressing over salad and toss lightly. Serves six.

TOSSED SPROUT SALAD

½ teaspoon salt
⅛ teaspoon pepper
¼ cup salad oil
1 tablespoon lemon juice
½ cup onion, finely chopped
1 clove garlic, cut in half

2 cups soy sprouts (page 38),
   cooked and chilled
1 head lettuce, torn in bite-size
   pieces
1 green pepper, coarsely
   chopped

Mix salt, pepper, oil, lemon juice, and 1 tablespoon onion. Chill. Rub wooden salad bowl with cut side of garlic clove. Toss soy sprouts, lettuce, green pepper, and remaining onion. Just before serving, pour chilled dressing over salad. Serves six.

SPROUT FRUIT SALAD

2 cups soy sprouts (page 38)
1 cup sunflower seeds
3 unpeeled apples, finely
   chopped

3 bananas, sliced
½ cup dates, chopped
1 cup soy yogurt (page 54)
1 tablespoon honey

Combine soy sprouts, seeds, and fruits. In another bowl, mix yogurt and honey. Pour yogurt over fruits. Serves six.

LOW-CALORIE
SOY DRESSING

1 cup soy yogurt (page 54)
2 tablespoons lemon juice
¼ teaspoon dry mustard
½ teaspoon salt

½ teaspoon paprika
1 clove garlic, minced
1 small onion, grated

Run all ingredients through blender until smooth. Chill.

CURD DRESSING

Blend soy curd (page 35) with mayonnaise or French dressing. Let stand 30 minutes before serving. If desired, add a cut clove of garlic.

CREAMY CURD DRESSING

½ cup fresh soy curd (page 35)   1 teaspoon sugar
1 tablespoon lemon juice   Soy milk (page 33)

Put curd, lemon juice, and sugar in blender. Blend, adding 1 teaspoon soy milk at a time, until smooth and creamy. Makes about 1 cup dressing.

YOGURT DRESSING

1 cup soy yogurt (page 54)   4 tablespoons Roquefort or
¼ teaspoon paprika   bleu cheese

Mix well in blender. Chill thoroughly.

Soy Spreads

That all-American lunch tradition, the sandwich, has graduated to a breakfast and dinner dish. Update your recipe files with some new ideas using protein-rich soybeans.

In addition to the soyloaf recipes we have given you which make good sandwich fillings, soybeans in almost every form may be used to make spreads for breads.

We'll start with two that have familiar flavors—Soy Margarine and Soy Nut Butter.

SOY MARGARINE

1 cup soy flour (page 29), sifted
1 tablespoon cornstarch
½ teaspoon salt
¾ cup water

1 tablespoon lemon juice
2 cups corn oil
2 or 3 drops yellow food
 coloring

Combine soy flour, cornstarch, and salt. Mix water and lemon juice. Gradually add to dry ingredients and stir to dissolve. Cook over hot water in the top of a double boiler until thick, stirring constantly. Empty into a crockery bowl. Add oil, ¼ cup at a time, to the hot mixture, beating thoroughly with an egg beater, electric mixer or blender after each addition. Repeat until all oil is used. Mix in enough food coloring to make mixture a light yellow color. Chill. Keep refrigerated. May be used in any way you would use butter or margarine.

SOY NUT BUTTER

Soy nuts (pages 47–48)
Salad oil

Salt to taste

Using fine blade of food grinder or blender, grind a cup of soy nuts at a time to a fine powder. A tablespoon at a time, add just enough oil to make a thick paste. Salt to taste. Cover and keep refrigerated.
Or:
Follow recipe above, using one half soy nuts and one half peanuts.
Or:
One tablespoon at a time, blend oil into roasted soy flour (page 49) to make a thick, smooth paste. Add roasted soy grits (page 48), to make crunchy soy nut butter.

CRUNCHY SOY NUT BUTTER

Add a few tablespoons coarsely ground soy nuts to first soy nut butter recipe, opposite.

SOYBEAN BACON SANDWICHES

3 tablespoons butter or soy margarine
6 slices bread
2 cups soy pulp (page 29)
¼ cup catsup

½ teaspoon prepared mustard
1 teaspoon brown sugar
6 slices American cheese
6 slices bacon, uncooked

Butter bread slices. In a bowl, combine soy pulp, catsup, mustard, and sugar. Blend well and spread on buttered bread slices. Top each slice with a slice of cheese and one slice of bacon cut in half. Broil until cheese melts and bacon is crisp. Makes six sandwiches.

SALMON-SOY SANDWICHES

1 (8-ounce) can salmon
3 hard-cooked eggs, chopped
¼ teaspoon Worcestershire sauce
2 tablespoons sweet pickle relish, drained

3 tablespoons mayonnaise or salad dressing
¼ cup roasted soy grits (page 48)
12 slices whole wheat bread
2 eggs, well beaten
⅔ cup soy milk (page 33)
Hot oil for frying

Drain, bone, and flake salmon. Combine with next five ingredients. Mix well. Spread on six slices of bread. Top with other six slices of bread. In a shallow bowl, combine beaten eggs and soy milk. Dip sandwiches in mixture and fry in hot oil until golden brown on both sides. Makes six sandwiches.

EGG-SPROUT FILLING

4 hard-cooked eggs, chopped
1 cup dried soy sprouts (page
   41) or roasted soy sprouts
   (page 49)

3 tablespoons ripe olives,
   chopped
¼ teaspoon salt
Mayonnaise or salad dressing

Combine first four ingredients. Moisten with mayonnaise or salad dressing. Enough for four sandwiches.

SOY-CHEESE SPREAD

1 cup soy curd (page 35)
1 cup Cheddar cheese, grated

¼ cup pimiento, finely diced
¼ cup sour cream

Combine all ingredients well. Refrigerate overnight to blend flavors. Enough for six sandwiches.

CURD-OLIVE SANDWICHES

1 cup soy curd (page 35)
¼ cup mayonnaise or salad
   dressing

¼ cup stuffed green olives,
   chopped
¼ cup dill pickle, chopped

Combine all ingredients and spread between slices of rye bread. Enough for six sandwiches.

CURD-SPROUT FILLING

½ cup soy curd (page 35)
2 tablespoons thick sour cream
½ teaspoon horseradish

½ teaspoon Worcestershire sauce
¼ cup roasted soy sprouts
   (pages 49–50), chopped

Combine all ingredients. Enough for four sandwiches.

CURD-ONION SANDWICHES

¾ cup soy curd
(page 35)
½ cup mayonnaise or salad
dressing

2 tablespoons pimiento, finely
diced
1 medium Bermuda onion,
thinly sliced

Combine soy curd, mayonnaise or salad dressing, and pimiento. Spread thickly on slices of whole wheat bread and top with onion slices. Top with buttered slice of bread. Enough for four sandwiches.

CURD-DRIED BEEF FILLING

1 cup soy curd (page 35)
¼ cup dried beef, finely shredded

1 teaspoon onion, grated

Blend all ingredients well. Enough for three sandwiches.

CURD-TURKEY SANDWICHES

2 tablespoons butter or
margarine
2 tablespoons flour
½ teaspoon salt
¼ teaspoon dry mustard
1 cup soy milk (page 33)
1 cup processed American
cheese, grated

6 slices toasted bread
6 slices pressed soy curd
(page 35)
12 slices cooked turkey
1 large tomato, thinly sliced
¼ teaspoon paprika

Melt margarine. Blend in flour, salt, and mustard. Gradually stir in soy milk and cook, stirring constantly, until mixture thickens. Add cheese and stir until cheese melts. Arrange toast slices on bottom of baking pan. Cover each slice with a slice of soy curd, two slices of turkey, and a slice of tomato. Pour cheese sauce over all and sprinkle with paprika. Broil two minutes or until lightly browned. Makes six servings.

Soy Snacks

Snacks of every variety—from seasoned soy nuts to an elegant party cheese ball—can be made from soybeans. Serve them for at-home snacks to add good nutrition to the family's diet or serve them for company. They're sure to make a hit with your guests for their delicious flavor. Only you will know you're saving money.

PARTY SOY BALLS

1 cup condensed cream of
   mushroom soup
2 cups cooked soy grits (page 28)
2 teaspoons chicken, beef or
   vegetable-flavored broth
   granules or powder
¼ cup dry bread crumbs

1 egg, slightly beaten
⅓ cup onion, finely diced
3 tablespoons bleu cheese,
   crumbled
1 tablespoon parsley, chopped
¼ cup soy yogurt (page 54)

Combine ¼ cup soup, soy grits, broth granules or powder, bread crumbs, egg, onion, and 1 tablespoon bleu cheese. Shape firmly into 1½-inch balls. Arrange in shallow baking pan and broil four inches from heat. Turn and brown on other side. Meanwhile, make sauce by combining parsley, remaining soup, and remaining cheese. Heat to boiling, stirring frequently. Stir in yogurt. Arrange soyballs in chafing dish and pour heated sauce over them. Serve with toothpicks. Makes about 50 soyballs.

SOYLOAF-CURD CANAPES

½ cup soy curd (page 35)
2 tablespoons dill pickle,
  chopped
2 tablespoons stuffed green
  olives, chopped

½ bologna-flavored soyloaf
  (page 45)
¼ cup French dressing
1 tablespoon Worcestershire
  sauce

Combine curd, pickles, and olives. Spread soyloaf slices with the mixture. Roll up and fasten with a toothpick. Dip in mixture of French dressing and Worcestershire sauce and broil until sizzling hot. Serves six.

CURD-CHEESE SNACKS

1 cup Cheddar cheese, grated
1 cup pressed soy curd (page 37),
  grated or diced fine
¾ cup fine dry bread crumbs
1 egg, separated

½ teaspoon prepared mustard
¼ teaspoon paprika
¼ teaspoon salt
Deep oil for frying (at least 1
  inch)

Combine cheese, curd, ¼ cup bread crumbs, egg yolk, mustard, paprika, and salt. Fold in stiffly beaten egg white. Form into balls about ¾-inch in diameter and roll in remaining ½ cup bread crumbs. Fry until golden brown in deep oil which has been heated to 375°. Makes about 36 balls.

Candy and Cookies

You can salve your conscience a little as you are whipping up candy and cookie treats for your family and friends. Soy products of every kind can find their way into sweets and give them status by adding positive food value. When Billy bites into your chocolate fudge, that crunch he'll enjoy will be from the soy nuts you have chopped and added.

COOKED CHOCOLATE FUDGE

2 squares (2 ounces)
  unsweetened chocolate,
  broken into small pieces
⅔ cup soy milk (page 33)
2 cups sugar
⅛ teaspoon salt

2 tablespoons butter or
  margarine
1 teaspoon vanilla
1 cup soy nuts (page 47),
  coarsely chopped

Heat chocolate and soy milk in medium saucepan, stirring constantly until chocolate is melted and smooth. Add sugar and salt and stir until sugar is dissolved and mixture comes to a boil. Lower heat and cook without stirring until a small amount dropped into cold water forms a soft ball (236°). Remove from heat. Cool slightly, then add butter or margarine and vanilla and allow to cool to lukewarm (110°) without stirring. Beat by hand or with electric mixer until thick. Add soy nuts and pour at once onto a greased plate or pan. Makes about one pound of fudge.

UNCOOKED SOY FUDGE

½ cup soy curd (page 35)
2 squares (2 ounces) unsweetened
  chocolate, melted
2 cups confectioner's sugar,
  sifted

½ teaspoon vanilla
⅛ teaspoon salt
½ cup soy nuts, coarsely
  chopped (pages 47–48)

Combine soy curd, chocolate, and confectioner's sugar. Add vanilla, salt, and chopped soy nuts. Mix well. Press into greased shallow pan and chill until firm, about 15 or 20 minutes. Cut into squares. Makes about one pound.

SOY NUT CANDY

½ cup soy flour (page 29)
2 cups brown sugar
¼ cup soy nut butter (page 100)

½ cup soy milk (page 33)
½ cup soy nuts (pages 47–48),
  coarsely chopped

Combine soy flour, brown sugar, soy nut butter, and soy milk. Slowly heat over low heat and cook five minutes, stirring constantly. Remove from heat, cool slightly, and beat until thick. Add soy nuts and pour into a greased, 8-inch square pan. When cold, cut into squares.

SOY NUT BRITTLE

1 cup soy nuts (pages 47–48)
2 cups granulated sugar

¼ teaspoon soda

Spread nuts over bottom of buttered platter or cookie sheet. Using a heavy iron skillet and stirring constantly, heat dry sugar until it melts and turns a light golden color. Sprinkle soda over top, stir in quickly and pour over soy nuts on platter. As soon as the brittle is cool enough to handle, invert pan to release brittle and pull and stretch as thin as possible. Break into pieces. Makes 1½ pounds.

SOY DIVINITY

2 cups sugar
½ cup hot water
½ cup light corn syrup
2 egg whites
Pinch of salt

4 tablespoons soy flour (page 29)
½ cup soy nuts (pages 47–48),
   coarsely ground
1 teaspoon vanilla

Combine sugar, water, and corn syrup in a large saucepan. Bring to a boil and cook over low heat until mixture reaches the hard ball stage (250°). Meanwhile, beat egg whites with a pinch of salt until they form stiff peaks. Pour hot syrup gradually into beaten egg whites, beating with an egg beater or electric mixer as syrup is added. Add soy flour, soy nuts, and vanilla and beat until stiff. Pour into greased pan to cool. Cut into squares. Makes about 1½ pounds.

SOY CRISPS

2 eggs
1 cup brown sugar
1 teaspoon vanilla

3 cups soy nuts (page 47),
   finely ground

Beat eggs until thick. Add brown sugar and vanilla. Blend in ground soy nuts to make a stiff dough. Drop by teaspoonfuls onto oiled cookie sheet. Mash to ¼-inch thickness with a fork dipped in water. Bake 8 to 10 minutes in 350° oven. Remove from sheet at once. Makes 4 dozen cookies.

FRUIT-SPROUT BALLS

1 cup soy sprouts (page 38)
½ cup soy curd (page 35)
2 cups soy nuts (page 47),
   coarsely ground

1 cup pitted dried fruit, chopped
   (prunes, dates, or apricots)

Combine sprouts, curd, 1 cup ground soy nuts, and chopped dried fruit. Shape into 1-inch balls and roll in remaining 1 cup ground soy nuts.

UNCOOKED FRUIT CANDY

1 cup pitted dates
½ cup seedless raisins
½ cup currants
1 cup soy nuts (pages 47–48)
1 cup soy nut butter (page 100)

2 tablespoons soy flour (page 29)
¼ cup soy milk (page 33)
Butter or margarine
Confectioner's sugar

Put fruits and nuts through coarse blade of food chopper. Add soy nut butter, soy flour, and enough soy milk to make a firm but sticky dough. Press into an 8 × 8 × 2-inch cake pan, which has been buttered, then sprinkled with confectioner's sugar. Smooth surface of candy with a knife dipped in hot water, then sprinkle top with confectioner's sugar. Chill until firm. Cut into squares. Makes about 1½ pounds.

SOY PULP COOKIES

3 tablespoons sugar
2 tablespoons butter or
  margarine
½ cup soy pulp (page 29)
1 teaspoon lemon juice

¼ teaspoon salt
¼ cup soy milk (page 33)
1 egg, well beaten
½ cup wheat flour
2 teaspoons baking powder

Cream sugar and butter or margarine. Combine soy pulp, lemon juice, and salt and add to creamed mixture. Add soy milk, egg, flour, and baking powder. Drop by teaspoonfuls onto greased cookie sheet. Bake 12 to 15 minutes in 350° oven. Makes two dozen.

SOY CRUNCHES

1 (8-ounce) package chocolate
  chips

2 cups soy nuts (page 47)

Melt chocolate in top of double boiler over hot water. Stir in soy nuts. Drop by spoonfuls on waxed paper and allow to harden. Makes 1½ pounds candy.

SOY COCONUT BALLS

¼ cup raisins
¼ cup pitted prunes
¼ cup dried apricots
¼ cup pitted dates
¼ cup soy nuts (page 47),
    coarsely ground

⅓ cup soy flour (page 29)
1 teaspoon grated orange rind
2 tablespoons orange juice
½ cup shredded coconut

Grind raisins, prunes, apricots, and dates through coarse blade of food chopper. Add soy nuts, soy flour, orange rind, and orange juice. Mix thoroughly and form into small balls. Roll in coconut. Makes about one pound of candy.

SOY SNAPS

1 cup solid vegetable shortening
1 cup molasses
1 egg, well beaten
1 teaspoon salt

1 teaspoon soda
3 teaspoons ginger
1 cup soy flour
3½ cups wheat flour

Melt shortening. Add molasses and egg and beat thoroughly. Sift remaining ingredients together and fold into mixture. Chill dough until firm enough to roll. Roll thin, cut with cookie cutter and bake in 350° oven until lightly browned. Makes two dozen.

SOY-OATMEAL COOKIES

½ cup solid vegetable shortening
½ cup butter or margarine
1 cup brown sugar
2 eggs, beaten
1 cup dried fruit (prunes,
    dates, or apricots), chopped

3 cups rolled oats
1 cup roasted soy flour (page 49)
1 cup wheat flour
¼ teaspoon salt
4 tablespoons soy milk (page 33)

Cream shortening, butter or margarine, and brown sugar until fluffy. Add eggs and dried fruit. Add oats, soy flour, wheat flour, and salt.

Add soy milk and mix well. Dough will be stiff. Drop by teaspoonfuls on greased cookie sheet. Flatten with a fork dipped in water and bake 8 to 10 minutes in 350° oven. Makes two dozen.

SPROUT COOKIE BARS

1 cup raisins
1 cup water
1 cup wheat germ
1 cup soy flour
2 teaspoons powdered cinnamon
¼ teaspoon powdered allspice
¼ teaspoon grated nutmeg
½ cup soy nuts (page 47),
   coarsely ground

1 cup soy sprouts, (page 38)
   chopped
¼ cup cooking or salad oil
2 tablespoons molasses
2 teaspoons vanilla
2 eggs, well beaten
Confectioner's sugar

Soak raisins overnight in water. Drain, reserving both raisins and water. Combine dry ingredients, then add soy nuts, soy sprouts, raisins, oil, molasses, vanilla, and ½ cup raisin-soaking water. Mix thoroughly. Add eggs. Pour into greased 9-inch square pan and bake 45 minutes at 350°. While still hot, cut into bars and roll in confectioner's sugar. Makes two dozen.

SOY SUGAR COOKIES

1 cup sugar
¾ cup solid vegetable shortening
2 eggs, well beaten
1 teaspoon vanilla
1 teaspoon salt

1 cup soy flour
1½ cups wheat flour
½ cup soy nuts (page 47),
   coarsely ground

Cream sugar and shortening until fluffy. Add eggs, vanilla, salt, soy flour, wheat flour, and soy nuts. Roll dough ¼-inch thick on a lightly floured board. Cut with a cookie cutter and place on greased cookie sheet. Bake in 400° oven 10 minutes, until lightly browned. Makes four to five dozen.

Soy Soups

Extra protein may be given to almost any soup by the addition of whole cooked soybeans, either fresh or dried, ground or mashed beans, soy milk, or soy flour.

Add uncooked soaked and ground soybeans to long-cooking soups, and leftover, cooked soybeans to vegetable soups. Although it does not have the thickening properties of wheat flour, soy flour may be blended into almost any but the clear soups to add protein and Vitamin B.

Use soy milk in any cream soup in place of cow's milk. Soy flour and water may be used as a quick substitute for soy milk in soups.

Fresh or cooked soy sprouts may be used as a crisp, nutritious garnish for almost any soup. They are especially delicious served in a clear meat or vegetable broth.

Pressed soy curd, which has little taste of its own, is a welcome addition to many soups. Cooked and sliced, curd is a nutritious, starch-free replacement for noodles or macaroni. Flavored and chopped, it can be added to vegetable soups in place of meat.

Anytime you get out the soup kettle, think of soybeans.

SPROUT SOUP

2 tablespoons butter or margarine
2½ cups soy sprouts (page 38)
5 cups beef stock or beef
  bouillon

Salt and pepper to taste
French bread
1 cup Parmesan cheese, grated

Melt butter or margarine in saucepan. Stirring constantly, sauté soy sprouts until lightly browned, about five minutes. Add stock or bouillon and cover. Lower heat and simmer 20 minutes. Season with salt and pepper. Place a slice of French bread in each soup bowl. Sprinkle with cheese and pour hot soup over bread. Serves eight.

ITALIAN SOYBEAN SOUP

¼ pound bacon, chopped
2 medium onions, sliced
1 clove garlic, minced
2 tablespoons cooking oil
½ cup celery, chopped
1 carrot, sliced
¼ cup parsley, chopped
1 cup cabbage, chopped
6 cups water

2 cups cooked dried soybeans
  (page 24)
2 cups cooked or canned
  tomatoes
½ teaspoon dried basil
¼ teaspoon oregano
2 teaspoons salt
1 pound hot Italian sausage,
  cut in ½-inch slices
½ cup grated Parmesan cheese

Brown bacon, onions, and garlic in oil in large pan or kettle. Add celery, carrot, parsley, and cabbage. Cook 10 minutes. Add water, soybeans, tomatoes, herbs, and salt. Cover and simmer 30 minutes. Top each serving with sausage and cheese. Serves eight.

SOY VEGETABLE SOUP

3 tablespoons butter or
  margarine
1 pound (about 2 cups) cooked
  soybeans (page 24), drained
3 cups cooked or canned
  tomatoes
4 cups soybean cooking liquid
  plus water
4 teaspoons vegetable broth
  granules or powder
1 package dry onion soup mix
4 carrots, sliced

1 cup fresh, canned, or
  frozen corn
½ cup celery tops, chopped
½ cup parsley, chopped
1 bay leaf
½ teaspoon dried oregano,
  crushed
½ cup uncooked macaroni
  alphabets
Salt and pepper
¼ cup grated Parmesan cheese

Melt butter or margarine in soup kettle. Add soybeans and cook, stirring, until lightly browned. Add next 10 ingredients. Simmer 30 minutes. Add macaroni alphabets and cook until tender. Season to taste with salt and pepper. Sprinkle each serving with Parmesan cheese. Serves six to eight.

SLOW COOKER SOUP

1½ cups dried soybeans   1 small onion
5 cups water   1 small ham shank
1 carrot   ½ teaspoon salt
½ rib celery   ⅛ teaspoon pepper

Soak soybeans in water 24 hours in refrigerator. Pour soybeans and all but 1 cup soaking water in slow cooker. Put remaining 1 cup water, carrot, celery, and onion quickly through blender. Add to slow cooker. Add ham, salt, and pepper. Stir to combine. Cover and cook 12 to 18 hours on low. Just before serving, remove ham. Discard fat, skin, and bone and cut meat into small pieces. Put one-half the cooked beans through the blender to make a puree. Return all to cooker and reheat. Serves four to six.

CURD-MUSHROOM SOUP

1 small onion, minced   ½ pound (about 1 cup) pressed
2 tablespoons butter or   or cooked soy curd (page 35),
    margarine   chopped fine
2 tablespoons wheat flour   ¼ pound fresh or canned
2 cups soy milk (page 33)   mushrooms, sliced
                                Salt and pepper to taste

Sauté onion until transparent in butter or margarine. Stir in flour, then gradually add soy milk. Cook, stirring constantly, over low heat until thickened. Add soy curd, mushrooms, and salt and pepper to taste. Serves four.

POTATO-CARROT CHOWDER

1 onion, chopped	1 teaspoon salt
¼ cup butter or margarine	¼ teaspoon paprika
2 cups raw potatoes, diced	2 tablespoons soy flour (page 29)
2 carrots, diced	1 tablespoon cornstarch
3 cups boiling water	2 cups soy milk (page 33)

Saute onion in 2 tablespoons butter or margarine in large saucepan. Add potatoes, carrots, boiling water, salt, and paprika. Cook, covered, about 30 minutes, until potatoes and carrots are tender. In a small bowl, blend soy flour and cornstarch into remaining butter or margarine. Add to potato mixture and cook, stirring, constantly. Gradually stir in milk. Cook five minutes, stirring, until smooth. Makes six servings.

Soymeat Dishes

Every time you push your supermarket cart up to the check-out counter, you brace yourself for the inevitable shock. *That much* for only two bags? Chances are the villain was your meat purchase.

Soybeans, with a little help from other meatless protein foods, can skirt the meat problem and give your wallet a breather now and then.

Invite Uncle Alistair over, after you have experimented with some of these ersatz meat dishes, and have a little fun fooling the gentleman with your thrifty but toothsome entrée.

SOYLOAF WITH
MUSHROOM SAUCE

1 beef-flavored soyloaf (page 44)
¼ cup wheat flour
3 tablespoons cooking or salad
  oil
1 medium onion, sliced

1 green pepper, cut in strips
1 can condensed cream of
  mushroom soup
½ cup soy milk (page 33)

Cut soyloaf in ½-inch slices. Dredge in flour and brown both sides quickly in hot oil, being careful not to break up slices when turning. Remove slices and add onion and green pepper to skillet. Cook, stirring constantly, two or three minutes, until lightly browned and mixed well. Spread out evenly over bottom of skillet and lay soyloaf slices over top. Combine soup and soy milk and pour over all. Bake in 350° oven 20 to 30 minutes. Serves four to six.

TAMALE LOAF

1 small onion, chopped
1 clove garlic, chopped
⅓ cup cooking or salad oil
1 beef-flavored soyloaf
  (page 44), chopped
1 teaspoon salt
3 cups cooked or canned
  tomatoes, chopped

2 teaspoons chili powder
Dash of cayenne
¾ cup yellow cornmeal
¾ cup soy milk (page 33)
1 cup canned cream-style corn
¾ cup pitted ripe olives,
  sliced

Sauté onion and garlic in oil five minutes. Add soyloaf and brown, stirring frequently. Add salt, tomatoes, chili powder, and cayenne. Simmer over low heat 10 minutes. Combine cornmeal with soy milk. Add to cooked mixture, mix well, and cook 15 minutes. Add corn and olives and pour into greased 8 × 12-inch pan. Bake one hour in 350° oven. Serves six.

SOYLOAF POT PIE

1 beef-flavored soyloaf (page 44), cut in ½-inch cubes
4 tablespoons wheat flour
2 tablespoons cooking or salad oil
2 cups cooked potatoes, cubed
8 small, cooked onions (about 1 inch in diameter)

½ cup canned or fresh mushrooms, sliced or cut in pieces
½ cup fresh or frozen peas
½ teaspoon salt
$1/8$ teaspoon pepper
2 teaspoons beef or vegetable broth granules or powder
2 cups hot water
1 teaspoon Worcestershire sauce

Dredge soyloaf cubes in 2 tablespoons wheat flour, then brown in hot oil. Combine in a casserole with potato cubes, onions, mushrooms, and peas. Sprinkle with salt and pepper. Combine remaining 2 tablespoons flour and broth granules or powder in a saucepan. Gradually add hot water and Worcestershire sauce. Cook, stirring constantly, until smooth and thickened. Pour over soyloaf and vegetables in casserole. Cover and bake 30 minutes in 350° oven. Serves four to six.

SOYLOAF CHILI

1 medium onion, chopped
2 tablespoons green pepper, chopped
1 clove garlic, minced
2 tablespoons cooking or salad oil
1 beef-flavored soyloaf (page 44), chopped

2 tablespoons chili powder
1 cup tomato juice
1 cup beef bouillon (or 1 bouillon cube dissolved in 1 cup water)
½ teaspoon salt
$1/8$ teaspoon pepper
2 cups cooked or canned pinto beans

Sauté onion, green pepper, and garlic in oil. Add chopped soyloaf and chili powder and cook two minutes, stirring constantly. Add tomato juice and beef bouillon and cook 20 minutes, stirring occasionally. Season to taste with salt and pepper. Add pinto beans and cook 20 minutes longer. Serves six.

MOCK STEAK

1 beef-flavored soyloaf (page
   44), coarsely chopped
½ cup quick-cooking rolled oats
   uncooked
½ teaspoon salt

¼ cup onion, finely chopped
2 tablespoons soy flour (page 29)
½ cup soy milk (page 33)
Butter or margarine

Mix together beef, oats, salt, and onion. Add soy flour to soy milk
and stir well until blended. Add to mixture. Pat on aluminum foil into
a flat, round "steak" about two inches thick. Broil four inches from
heat about 20 minutes, or until browned. Dot surface with butter or
margarine. Serves six.

STUFFED PEPPERS

6 medium green peppers
4 cups boiling water
1 medium onion, chopped
1 tablespoon cooking or salad oil
1 beef-flavored soyloaf (page
   44), chopped
1 cup soft bread crumbs
1 cup canned or fresh tomatoes,
   chopped

¼ cup celery, chopped
½ teaspoon salt
⅛ teaspoon pepper
½ cup dry bread crumbs
2 tablespoons butter or
   margarine, melted
2 tablespoons grated Parmesan
   cheese

Cut off tops of peppers and remove core and seeds. Drop into boiling
water, cover, and remove from heat. Let set 10 minutes. Meanwhile,
saute onion in oil until lightly browned. Add chopped soyloaf, soft
bread crumbs, tomato, celery, and salt and pepper. Remove peppers
from water and drain, upside down. Fill with soyloaf mixture and
top with crumb topping made by combining dry bread crumbs,
melted butter or margarine, and Parmesan cheese. Bake 45 minutes
in 350° oven. Makes six servings.

SOYLOAF HASH

2 cups uncooked potatoes,
  peeled and cut in ½-inch
  cubes
½ cup onion, chopped
1 cup water
1 teaspoon beef broth granules
  or powder

2 tablespoons butter or
  margarine
1 beef-flavored soyloaf (page
  44), cut in ½-inch cubes
Salt and pepper

Cook potatoes and onion in water in tightly covered saucepan 20 minutes, or until tender. Add broth concentrate and butter or margarine. Stir until dissolved. Add soyloaf cubes and season to taste with salt and pepper. Pour into baking dish and bake, uncovered, 30 minutes, in 350° oven, until heated through and browned on top. Serves six.

SOYLOAF-CHEESE DINNER

1½ cups uncooked spaghetti,
  broken into 1-inch pieces
3 cups boiling water
½ teaspoon salt
½ beef-flavored soyloaf
  (page 44), chopped
1 small onion, chopped

2 tablespoons butter or
  margarine
3 tablespoons wheat flour
2 cups soy milk (page 33)
Salt and pepper
½ cup soy curd (page 35)
½ cup Cheddar cheese, grated

Cook spaghetti until tender in 3 cups boiling water to which ½ teaspoon salt has been added. Drain, discarding water. In a skillet, brown chopped soyloaf and onion in butter or margarine. Blend in flour, then gradually add soy milk. Cook, stirring constantly, until thickened. Season to taste with salt and pepper. In a bowl, combine curd and cheese. Stir one-half of this mixture into the cooked spaghetti. Place one-half the cooked spaghetti mixture in a greased casserole. Pour beef mixture over this. Add the tomatoes, then the remaining spaghetti mixture. Sprinkle with remaining cheese mixture. Bake 25 to 30 minutes in a 350° oven. Serves six to eight.

SOYLOAF MEAT LOAF

1 beef-flavored soyloaf
  (page 44), chopped
½ cup rolled oats
½ cup cracker crumbs
½ cup soy flour (page 29)
1 teaspoon salt
$^1/_8$ teaspoon pepper
1 can condensed tomato soup,
  undiluted

2 teaspoons beef broth granules
  or powder
1 teaspoon Worcestershire sauce
1 tablespoon Kitchen Bouquet
½ cup uncooked carrot, grated
¼ cup celery, finely chopped
1 small onion, chopped
1 egg
¼ cup catsup

Combine all ingredients except catsup. Pack into a greased 8 × 5-inch loaf pan and bake 45 minutes in a 350° oven. Spread catsup over top of loaf and return to oven 15 minutes more. Cool 10 minutes before slicing. Serves six.

SPIRALED SOYLOAF

Filling:

½ chicken-flavored soyloaf
  (page 45), diced
½ can condensed cream of
  mushroom soup
1 cup celery, chopped

¼ cup green pepper, chopped
1 tablespoon onion, chopped
3 hard-cooked eggs, sliced
½ teaspoon salt

Combine all ingredients and mix well.

Pastry:

1¼ cups wheat flour
¼ cup soy flour (page 29)
½ cup cornmeal
1 teaspoon baking powder

1 teaspoon salt
½ cup solid vegetable
  shortening
½ cup soy milk (page 33)

Sift together dry ingredients. Cut in shortening with a pastry blender or two knives until mixture resembles coarse meal. Add soy milk, stirring with a fork until dough forms a ball. Knead gently a few times on a lightly floured board. Roll out to a 10 × 12-inch rectangle. Spread filling evenly over dough. Roll up as for a jelly roll, sealing the edge. Bake 20 to 25 minutes in a 425° oven. Slice and serve with sauce made by adding ½ cup soy milk to the remaining ½ cup mushroom soup. Heat to boiling. Serves six.

TEXAS SOYLOAF STEW

1 beef-flavored soyloaf (page 44), chopped
2 medium onions, chopped
2 cloves garlic, minced
2 tablespoons cooking or salad oil
2 tablespoons chili powder

2 tablespoons wheat flour
2 cups cooked or canned tomatoes, chopped
2 cups cooked soybeans (page 24)
2 bay leaves
Salt and pepper

Brown soyloaf, onions, and garlic in oil. Stir in chili powder which has been mixed with flour. Add remaining ingredients and simmer 45 minutes. Remove bay leaves before serving. Serves six.

SOY SUEY

½ cup onion, chopped
2 tablespoons cooking or salad oil
1 cup celery, sliced
½ cup green pepper, chopped
2 cups soy sprouts (page 38)
3 teaspoons soy sauce
1 cup boiling water

1 teaspoon chicken broth granules or powder
2 tablespoons cornstarch
2 tablespoons cold water
½ chicken-flavored soyloaf (page 45), cut in thin strips, 1½ inches long

Sauté onion in oil until transparent. Add celery, green pepper, sprouts, soy sauce, and 1 cup boiling water in which broth granules or powder have been dissolved. Heat to simmering. Cover, lower heat, and cook 10 to 15 minutes, just until celery is tender-crisp. Dissolve cornstarch in cold water and add. Cook, stirring constantly, until thickened. Stir in soyloaf strips and cook over low heat 10 minutes longer. Serve over hot, cooked rice. Serves six.

SOY-POTATO CASSEROLE

1 bologna-flavored soyloaf
  (page 45)
4 large potatoes, sliced
¼ cup wheat flour
¼ cup soy flour (page 29)
½ teaspoon salt

⅛ teaspoon pepper
2 cups soy milk (page 33),
  heated to boiling
¼ pound American processed
  cheese, grated

Cut soyloaf into slices. Set aside. Spread one-half the potato slices in bottom of a greased shallow casserole. Combine wheat flour, salt, and pepper and sprinkle over potatoes. Top with soyloaf slices, then with remainder of potato slices. Pour hot soy milk over all. Bake 30 minutes in 350° oven. Top with grated cheese, then return to oven 15 minutes more. Serves six.

SOYLOAF DRUMSTICKS

3 tablespoons butter or
  margarine
3 tablespoons wheat flour
1 tablespoon soy flour (page 29)
1 cup soy milk (page 33)
2 eggs
1 teaspoon Worcestershire sauce
2 tablespoons parsley, chopped

2 tablespoons onion, minced
½ teaspoon salt
2 drops tabasco sauce
1 chicken-flavored soyloaf (page
  45), chopped
2 tablespoons water
1 cup fine cracker crumbs
Hot oil for frying

Melt butter or margarine. Blend in wheat flour and soy flour. Gradually stir in soy milk and cook over low heat until very thick. Beat one of the eggs and add to sauce. Stir in Worcestershire sauce, parsley, onion, salt, tabasco, and chopped soyloaf. Mix well and chill. Shape into drumsticks. In a bowl, beat remaining egg. Add water and dip "drumsticks" into egg mixture, then roll in cracker crumbs. Fry until golden brown in deep oil heated to 375°. Drain on paper towels and insert wooden skewers in ends. Serves four.

SOYLOAF CROQUETTES

1 medium onion, chopped
2 tablespoons cooking or salad
  oil
2 cups soy sprouts (page 38),
  chopped and drained

1 chicken-flavored soyloaf
  (page 45),
¼ cup mayonnaise
2 eggs, beaten
1 cup corn flakes, crushed
Hot oil for frying

Sauté chopped onion in oil. Add soy sprouts and stir-fry two minutes. Combine with soyloaf. Add mayonnaise and mix lightly. Shape into croquettes and chill two hours. Dip in beaten eggs, then in corn flake crumbs; again in eggs and in crumbs. Fry in hot oil until golden brown. Serves six.

HAWAIIAN SOYLOAF

1 egg, slightly beaten
2 tablespoons water
¼ cup soy flour (page 29)
1/8 teaspoon salt
1 chicken-flavored soyloaf (page
  45), cut into 1-inch cubes
Hot oil for frying
1 tablespoon butter or
  margarine
½ cup pineapple juice

1½ cups water
1 chicken-flavored bouillon
  cube
1 large carrot, thinly sliced
1 medium green pepper, sliced
  lengthwise
1 cup pineapple chunks
2 tablespoons cornstarch
1 tablespoon soy sauce
1 tablespoon water

Put egg in bowl. Add water, soy flour, and salt. Beat until smooth. Add soyloaf cubes and stir until cubes are coated. Drop into hot oil and fry until browned on all sides. Melt butter or margarine. Add pineapple juice, 1½ cups water and bouillon cube. Stir to dissolve bouillon. Add carrot. Cover and cook five minutes over low heat. Add green pepper and pineapple chunks. Dissolve cornstarch in soy sauce and 1 tablespoon water. Stir into mixture and cook, stirring constantly, until thickened. Add soyloaf cubes and serve over hot, cooked rice. Serves four.

SOY SAUSAGE

2 cups cooked soy grits (page 28)
2 tablespoons cooking or salad oil
½ cup wheat flour
½ cup soft bread crumbs
1 egg

1 teaspoon salt
¼ teaspoon pepper
⅛ teaspoon cayenne
1½ teaspoons dried sage, crushed fine
Oil for frying

Combine all ingredients. Shape into patties. Heat oil in heavy skillet and fry patties just until browned on both sides or brown in broiler. Serves six.

SOY SAUSAGE CASSEROLE

1 cup dry soybeans
3 cups water
1 teaspoon salt
2 tablespoons cooking or salad oil
3 tablespoons butter or margarine
2 tablespoons wheat flour

1 cup soy milk (page 33)
¼ teaspoon salt
⅛ teaspoon pepper
½ cup soy curd (page 35)
½ cup Cheddar cheese, grated
½ soy sausage (previous recipe) cooked, then crumbled
1 cup bread crumbs

Cook soybeans in water two minutes. Let set one hour or longer, then add salt and 1 tablespoon oil and cook two hours, or until tender. Meanwhile, melt two tablespoons butter or margarine and blend in flour. Gradually stir in soy milk. Cook, stirring constantly, until thickened. Season with salt and pepper. Add curd and cheese and stir to blend well. Add drained, cooked soybeans and soy sausage. Pour into casserole. Melt remaining 1 tablespoon butter or margarine and stir in bread crumbs. Sprinkle crumbs over top of casserole. Bake in 350° oven about 20 minutes, until crumbs are browned. Serves four.

STUFFED TOMATOES

6 medium tomatoes
1 tablespoon onion, chopped
1 tablespoon green pepper,
  chopped
1 tablespoon cooking or salad oil
1 cup soy sausage, (page 124),
  cooked and crumbled

½ cup toasted bread crumbs
½ teaspoon salt
⅛ teaspoon pepper
3 tablespoons grated Parmesan
  cheese

Remove tops and scoop out centers from tomatoes, reserving pulp. Sauté onion and green pepper in oil until onion is transparent. Combine with soy sausage, bread crumbs, salt, pepper, and tomato pulp. Fill tomato shells with mixture and top with Parmesan cheese. Bake 25 minutes in 350° oven. Serves six.

GRILLED SOY DOGS

2 cups cooked soy grits (page 28)
½ cup soy flour (page 29)
½ cup wheat flour
¼ cup cooking or salad oil
2 teaspoons liquid smoke
  flavoring or hickory smoked
  salt
2 teaspoons salt (reduce to 1
  teaspoon if hickory smoked salt
  is used)

Water
Dill pickles, quartered
  lengthwise
Barbecue sauce
Processed cheese slices
Hot dog buns

Combine first six ingredients. Stir in enough water to make a workable dough. For each soy dog, form a piece of dough around a dill pickle quarter in the shape of a hot dog. Broil in the oven or on an outdoor grill, basting with barbecue sauce and turning to brown on all sides. Top with a strip of cheese cut to fit and return to broiler or grill until cheese melts. Serve on hot dog buns. Serves four to six.

LENTIL-SOY NUT ROAST

½ cup soy nuts (pages 47–48)
1 cup cooked lentils
1 egg
1 large (13 ounce) can
  evaporated milk
½ cup salad oil

1½ cups soy-wheat cereal
  (page 71)
½ teaspoon dried sage,
  crumbled
2 tablespoons onion, minced
1 teaspoon salt

Put soy nuts through food grinder or blender to grind medium fine.
Combine with remaining ingredients. Shape into the form of a roast
and put in a greased baking dish. Bake 45 minutes in a 350° oven.
Makes four servings.

SOY-MUSHROOM BURGERS

½ pound fresh or canned
  mushrooms, sliced
3 tablespoons butter or
  margarine
2 cups cooked soy grits (page 28)
1 teaspoon onion juice

1 tablespoon Worcestershire
  sauce
1 tablespoon soy sauce
1 teaspoon salt
⅛ teaspoon pepper
4 hamburger buns

Sauté mushrooms in 2 tablespoons butter until lightly browned. Add
cooked soy grits and cook, stirring constantly, five minutes. Add onion
juice and seasonings. Spread each bottom half of hamburger bun with
a quarter of the mixture. Dot with remaining butter or margarine and
broil until lightly browned. Cover with top half of bun. Makes four
servings.

SOY BURGERS

2 cups cooked soy grits (page 28)
1 cup wheat germ
2 eggs, well beaten
½ cup whole wheat flour

6 hamburger buns
3 tablespoons butter or
  margarine

Combine cooked grits, wheat germ, and eggs. Shape into six patties. Coat lightly with whole wheat flour and brown on both sides in melted butter or margarine. Turn carefully to keep patties from breaking up. Serve on buns with pickle, onion and catsup. Serves six.

SOYBEAN CHILIBURGERS

1 medium onion, sliced
1 clove garlic, minced
2 tablespoons cooking or salad oil
3 cups cooked dried soybeans (page 24)

½ cup canned tomato sauce
¾ teaspoon chili powder
½ teaspoon salt
6 slices toast
½ cup Cheddar cheese, grated

Cook onion and garlic in oil until transparent. Add soybeans, tomato sauce, chili powder, and salt. Simmer 30 minutes, stirring frequently to keep from sticking. Spread over slices of toast and top with cheese. Broil until lightly browned. Makes six servings.

SLOPPY SOYBURGERS

2 tablespoons cooking or salad oil
1 green pepper, finely chopped
1 small onion, chopped
2 cups cooked soy grits (page 28)
2 cups cooked or canned tomatoes
¼ cup catsup

2 teaspoons Worcestershire sauce
2 teaspoons beef broth granules or powder
1 teaspoon salt (omit if broth is salted)
1 tablespoon cornstarch
1 tablespoon water
6 hamburger buns

Heat oil in skillet. Add green pepper and onion and cook until onion is tender. Add soy grits and cook, stirring constantly, until lightly browned. Add tomatoes, catsup, Worcestershire sauce, broth granules, and salt. Cook, uncovered, 15 minutes, stirring frequently. Meanwhile, blend cornstarch and water in a cup. Stir into tomato mixture and cook, stirring constantly, until thick. Serve on buns. Serves six.

SOY NUT LOAF

1½ cups soy nuts (page 47),
   coarsely ground
2 cups dry, whole wheat bread
  crumbs
2 tablespoons green pepper,
  finely chopped
2 tablespoons onion, finely
  chopped
1 tablespoon Worcestershire sauce

1 egg
½ cup cooked or canned
   tomatoes, run through
   blender
1 teaspoon salt (omit if soy nuts
   are salted)
½ teaspoon celery salt

Combine all ingredients, adding water if needed to moisten. Form into a loaf and let stand 10 minutes. Turn into a greased loaf pan and bake one hour in a 350° oven. Serves four to six.

SOY NUT CUTLETS

2 eggs, well beaten
1 tablespoon water
1 cup soy nuts (page 47),
  coarsely ground
1 cup soy-wheat cereal (page 71)
½ cup soy milk (page 33)

1 teaspoon lemon juice
1 small onion, chopped
2 tablespoons cooking or salad
  oil
1 cup cracker crumbs
1 inch deep oil for frying

Blend eggs and water. Combine one-half this mixture with soy nuts, cereal, soy milk, and lemon juice. Set aside. Cook chopped onion in oil until onion is transparent. Add to mixture and shape into four cutlet-shaped servings. Carefully dip the cutlets in reserved egg mixture, then in cracker crumbs. Fry in hot, deep oil until golden brown. Serves four.

SOY CURD FRITTERS

1 cup soy curd (page 35)
1 egg, well beaten
¼ cup rich chicken broth
¾ cup wheat flour

¼ cup soy flour (page 29)
2 teaspoons baking powder
½ teaspoon salt

Add soy curd to egg and beat until thoroughly blended. Stir in chicken broth. Mix wheat flour, soy flour, baking powder, and salt and add to first mixture and stir in lightly. Drop by spoonfuls into deep hot fat (375°) and fry until golden, two to four minutes. Drain on paper toweling. Makes 10 to 12 fritters.

SOY CURD CUTLETS

1 pound pressed soy curd
  (page 35), chopped
2 cups soft bread crumbs
½ cup soy nuts (pages 47–48),
  coarsely ground
2 tablespoons onion, minced
2 tablespoons green pepper,
  minced

¼ teaspoon paprika
1 teaspoon salt
1 egg
2 tablespoons soy milk (page 33)
½ cup cracker crumbs
¼ cup butter or margarine

Combine curd, crumbs, soy nuts, onion, green pepper, and seasonings. Shape into cutlets. Mix egg and soy milk and dip each cutlet into the mixture, then in cracker crumbs. Brown in hot butter or margarine. Serves four.

SOY PULP PATTIES

1 tablespoon onion, chopped
2 tablespoons butter or
  margarine
3 tablespoons wheat flour
⅓ cup soy milk (page 33)

1 egg, slightly beaten
2 cups soy pulp (page 29)
1 teaspoon salt
½ teaspoon dried sage,
  crumbled

Sauté onion in butter or margarine until transparent and golden. Stir in flour and gradually add soy milk. Cook over low heat, stirring constantly, until thickened. Add the egg and cook until very thick. Add soy pulp, salt, and sage. Stir well, then chill. Shape into patties about 3½ inches in diameter and broil until browned on one side. Turn and brown on other side. Serves four.

SOY HEALTHBURGERS

¼ cup soy grits (page 27)
½ cup meat stock or beef or
vegetable broth
2 cups soy pulp (page 29)
1 small onion, grated
1 clove garlic, minced
1 carrot, grated
1 rib celery, chopped

½ cup wheat germ
2 eggs, well beaten
3 tablespoons cooking or salad
oil
3 tablespoons nutritional yeast
½ teaspoon salt
½ cup cooked brown rice
2 sprigs fresh dill, minced

Soak soy grits in meat stock or broth overnight. Add remaining ingredients
and blend well. Mold into patties and broil until brown on both sides.
Serves six.

SOYBEAN-CURD ROAST

1 small onion, minced
2 tablespoons cooking or salad
oil
1 cup cooked brown rice
1 cup soy pulp (page 29)
1 cup soy curd (page 35)
1 cup soy-wheat cereal (page 71)

1 tablespoon vegetable broth
granules or powder
½ cup tomato purée
½ teaspoon celery salt
½ teaspoon salt
⅛ teaspoon pepper

Sauté onion until transparent in oil. Combine with other ingredients
and mix well. Shape into a roast and bake in a shallow pan 45 minutes
in a 350° oven. Makes six servings.

Soybeans With Meat

Perhaps the mayor and his wife are coming to dinner and you just have to trot out your best party manners—and at least *some* meat seems called for on the menu. Serve Stuffed Cabbage Rolls and see if His Honor doesn't smile appropriately.

COMPANY BAKED SOYBEANS

6 cups cooked dried soybeans (page 24), drained
2 cups cooked or canned fresh soybeans (page 22), drained
2 cups cooked or canned kidney beans, drained
2 cups cooked or canned pinto beans, drained
½ cup liquid drained from beans (cook all drained liquid down to ½ cup)

1 pound link sausage (Italian or smoked)
½ pound ham, cut in ½-inch cubes
1 teaspoon salt
½ teaspoon dry mustard
½ teaspoon pepper
1 cup canned tomato sauce
½ cup catsup
½ cup brown sugar
1 medium onion

Combine four kinds of beans and liquid in a large casserole. Cut sausage links into one-inch pieces and brown in heavy skillet. Drain off fat and add ham, salt, mustard, pepper, tomato sauce, catsup, and brown sugar. Mix well and pour over beans. Insert onion in the center of the beans. Bake one hour in 400° oven. Makes twelve servings.

NEW ENGLAND
BAKED SOYBEANS

2 cups dried soybeans
6 cups water
1 teaspoon salt
¼ pound salt pork, diced

½ cup onion, chopped
¼ cup brown sugar, packed
¼ cup molasses
1 teaspoon dry mustard

Soak soybeans 24 hours in refrigerator in 6 cups water. Add salt and cook in pressure pan 30 minutes at 15 pounds pressure. Drain, reserving cooking liquid. Pour soybeans into baking dish or beanpot. Combine remaining ingredients and ¾ cup of the reserved cooking liquid. Stir into beans. Cover casserole and bake three hours in a 325° oven. Remove cover and bake one hour longer. Makes six servings.

SOYBEANS, COUNTRY STYLE

1 pound dried soybeans
6 cups cold water
2 teaspoons salt
¼ pound pork sausage
1 medium onion, coarsely
  chopped
2 large tart apples, peeled and
  thinly sliced

2 cloves garlic, minced
⅛ teaspoon pepper
1 teaspoon chili powder
1 teaspoon dry mustard
¼ cup brown sugar, firmly
  packed
1½ cups tomato juice
½ cup sour cream

Soak soybeans 24 hours in refrigerator in 6 cups water. Add salt and cook in soaking water over low heat until beans are tender, adding water as necessary. Meanwhile, crumble sausage and brown in a skillet. Add onion, apple slices, and garlic and brown quickly, stirring constantly. Add all ingredients except sour cream to beans and cook over low heat two more hours. Place a spoonful of sour cream on each serving. Serves eight to ten.

LAYERED SOYBEAN CASSEROLE

2 tablespoons cooking or salad
  oil
1 pound ground beef
2 cups soy pulp (page 29)
1 clove garlic, minced
4 cups raw potatoes, sliced
1 pound fresh soybeans,
  uncooked (or 2 cups canned
  fresh soybeans, drained)
12 small white onions (or 1 can
  small white onions, drained)
1 teaspoon salt

⅛ teaspoon pepper
¼ teaspoon thyme
1 tablespoon cornstarch
1 cup tomato sauce
1 beef bouillon cube (or 1
  teaspoon beef broth granules
  or powder)
⅔ cup boiling water
½ cup Cheddar cheese, grated
¼ cup roasted soy grits (page 48)
3 tablespoons butter or
  margarine, melted

Heat oil in heavy skillet. Combine ground beef and soy pulp and
sprinkle with minced garlic. Sauté in oil until mixture is browned and
crumbly. Arrange layers in a greased three-quart casserole as follows:
One-half the potatoes, one-half the meat mixture, one-half the fresh
soybeans, all the onions, one-half the fresh soybeans, one-half the meat,
one-half the potatoes. Season each layer with salt, pepper, and thyme.
In a small bowl, dissolve cornstarch in 1 tablespoon tomato sauce. In
a cup, dissolve bouillon cube or broth granules in boiling water.
Combine both mixtures and add remaining tomato sauce. Pour all
over casserole. Combine grated cheese and roasted soy grits and spread
over casserole. Top with melted butter or margarine. Bake, uncovered,
in 350° oven 75 minutes, or until potatoes are tender and top is golden
brown. Serves six to eight.

STUFFED CABBAGE ROLLS

¾ pound ground beef
1 cup cooked soy grits (page 28)
1½ teaspoons salt
⅛ teaspoon pepper
½ cup onion, grated

½ cup raw rice
1 head cabbage
1 beef bouillon cube or 1
    teaspoon beef broth granules
1 cup boiling water

Combine first six ingredients. Remove core from cabbage and place head in boiling water for a few minutes to wilt leaves. Separate leaves. Put a heaping tablespoon of meat-rice mixture in the middle of a cabbage leaf. Roll up leaf and fasten with a toothpick. Place roll in casserole or baking pan. Repeat until all meat mixture is used. Dissolve bouillon cube or broth granules in boiling water. Pour over cabbage rolls. Cover and bake in 350° oven 1½ hours, or until rice is tender. Serves six.

SWEDISH MEAT-SOY BALLS

½ pound ground beef
1 cup cooked soy grits
    (page 28)
1 cup dry bread crumbs
1 teaspoon cornstarch
1 teaspoon salt
¼ teaspoon pepper
⅛ teaspoon powdered allspice

1 egg, beaten
1 cup soy milk (page 33)
1 small onion, minced
3 tablespoons cooking or salad
    oil
3 tablespoons wheat flour
2 cups water
⅔ cup red wine

Combine beef, soy grits, bread crumbs, cornstarch, salt, pepper, allspice, egg, and soy milk. In a skillet, saute onion in 1 tablespoon oil. Add to beef mixture. Blend well. Shape into about 40 tiny balls. Brown in remaining 2 tablespoons oil. Remove from pan. Make gravy by blending the flour into the oil in the skillet. Slowly add water and wine, stirring constantly. Season to taste with salt and pepper. Put meatballs in gravy and simmer 20 minutes. Serves six.

CORN PONE SOY PIE

½ pound ground beef
1 cup uncooked soy grits
  (page 27)
1 medium onion, chopped
1 tablespoon cooking or salad
  oil
2 teaspoons chili powder

¾ teaspoon salt
1 teaspoon Worcestershire sauce
1 cup cooked or canned
  tomatoes
1 cup cooked soybeans (page
  24), drained
Corn pone batter (below)

Brown ground beef, soy grits, and onion in oil. Add seasonings and tomatoes and simmer over low heat 20 minutes. Add soybeans and heat five minutes more. Pour into greased casserole, top with corn pone batter and bake 20 minutes in 400° oven or until golden brown. Serves six.

Corn Pone Batter:

1 cup yellow cornmeal
¼ cup soy flour (page 29)
¼ teaspoon salt
1 teaspoon baking powder

1 egg, well beaten
½ cup soy milk (page 33)
1 tablespoon cooking or salad
  oil

Combine dry ingredients. In another bowl, mix egg, soy milk, and oil. Add, all at once, to dry ingredients. Mix just until blended.

SOYBEANS AND RICE

4 slices bacon, diced
1 medium onion, diced
1 green pepper, diced
1 cup tomato sauce
½ teaspoon powdered oregano

¼ teaspoon garlic powder
4 cups cooked soybeans (page
  24), drained
2 cups hot, cooked rice

Fry bacon until crisp. Add onion and green pepper and cook over low heat until transparent, about 10 minutes. Add tomato sauce, oregano, and garlic powder. Simmer 10 minutes over low heat. Add soybeans and 1 cup cooking liquid. Cover and simmer 15 minutes. Serve over hot cooked rice. Serves six.

HAMBURGER-SOY PIE

½ pound ground beef
1 cup cooked soy grits (page 28)
1 cup onion, chopped
1 tablespoon cooking or salad
  oil
1 (10½-ounce) can condensed
  tomato soup

1½ teaspoons salt
1 cup cooked fresh soybeans
  (page 22), drained
1 egg
2 tablespoons butter or
  margarine
3 cups cooked potatoes, mashed

Stir and cook ground beef, grits, and onion in hot oil in heavy skillet until browned. Add soup, 1 teaspoon salt and fresh soybeans. Pour into 1½-quart casserole. In another bowl, combine egg, butter or margarine, remaining ½ teaspoon salt with mashed potatoes. Spread over top of casserole. Bake 30 minutes in 350° oven. Makes six servings.

CHINESE MEAT BALLS

1 teaspoon salt
½ pound ground beef
1 cup soy pulp (page 29)
3 tablespoons cooking or salad
  oil
¼ cup onion, chopped
1 cup celery, diced
1 cup carrots, thinly sliced

1 cup soy sprouts (page 38)
2 cups cooked fresh soybeans
  (page 22), drained
1 cup beef broth
2 tablespoons cornstarch
1 tablespoon soy sauce
¼ cup water
3 cups cooked rice

Add salt to ground beef and soy pulp. Combine. Shape into small balls and brown in oil. Move to one side. Add onion to skillet and cook, stirring constantly, until onion is transparent. Add celery, carrots, soy sprouts, fresh soybeans, and beef broth. Stir to combine, cover and cook 15 minutes. In a cup mix cornstarch, soy sauce, and water. Add to hot mixture and cook, stirring constantly, until thickened. Serve over cooked rice. Makes six servings.

BARBECUED MEAT BALLS

½ pound ground beef
1 cup cooked soy grits
  (page 28)
1 teaspoon salt
⅛ teaspoon pepper
⅓ cup fine, dry bread
  crumbs
¾ cup water
2 tablespoons cooking or salad
  oil

½ cup catsup
2 tablespoons brown sugar
2 tablespoons vinegar
2 teaspoons Worcestershire
  sauce
1 teaspoon prepared mustard
4 cups cooked soybeans (page
  24) drained
½ cup cooking water from
  soybeans

Combine beef, soy grits, salt, pepper, crumbs, and water. Form into 1½-inch balls. Brown on all sides in hot oil. Add remaining ingredients and put into a 1½-quart casserole. Bake one hour in a 350° oven. Serves six.

ORIENTAL PATTIES

6 slices bacon, diced
1 large onion, chopped
6 eggs, well beaten
2 cups soy sprouts (page 38)

1 cup cooked shrimp, chopped
Salt and pepper
Hot oil for frying

Cook bacon and onion until bacon is crisp and onion is transparent. Drain. In a bowl, combine eggs, soy sprouts, and shrimp. Add bacon and onions and blend. Using a ¼-cup measure to pour batter, brown patties in a hot, well-oiled skillet. Serves six.

TOMATO MEAT BALLS

1 pound ground beef
1 egg
1 medium onion, chopped
2 tablespoons parsley, chopped
1 teaspoon salt
¼ teaspoon basil

⅛ teaspoon pepper
½ cup cooked soy grits (page 28)
3 cups tomato juice
1 cup celery, thinly sliced
½ teaspoon chili powder
½ teaspoon salt

Combine beef, egg, onion, parsley, 1 teaspoon salt, basil, pepper, and soy grits. Shape into 12 small balls. In a large skillet, bring tomato juice, celery, chili powder, and ½ teaspoon salt to a boil. Drop in meatballs and lower heat. Cover and simmer 40 minutes. Serves six.

Meat Loaves

SOY SPROUT MEAT LOAF

1 pound ground beef
½ pound ground pork
1½ cups soy sprouts (page 38)
2 eggs, beaten
1 medium onion, quartered
1 medium carrot
½ green pepper
1 rib celery

1 cup tomato (or other
   vegetable) juice
½ cup rolled oats, uncooked
1 cup cracker or bread crumbs
1 teaspoon salt
1 tablespoon *Kitchen Bouquet*
   or Worcestershire sauce

Put beef and pork in large mixing bowl. In an electric blender, purée soy sprouts, eggs, onion, carrot, green pepper, celery, and tomato juice. Pour over meats. Add oats, crumbs, salt, and seasoning. Mix to blend well and shape into a large meat loaf or pack into a loaf pan. Bake one hour in a 350° oven. Serves six to eight.

SOY-BEEF LOAF

1 pound ground beef
1 cup soy pulp (page 29)
2 tablespoons onion, minced
2 tablespoons green pepper,
  finely chopped
2 tablespoons celery, finely
  chopped

2 eggs
1 teaspoon salt
¼ cup catsup
1 teaspoon Worcestershire sauce
1 cup cracker or bread crumbs

Combine all ingredients, mixing well. Press into a 9 × 5-inch loaf pan. Bake 75 minutes in a 350° oven. Cool 10 minutes. Run knife around edges of pan and invert onto platter. Cut into slices. Serves eight.

SPICY MEAT LOAF

1 pound ground beef
½ pound pork sausage
¾ cup cooked soy grits (page 28)
1 medium onion, finely chopped
1 egg
1 cup soft bread crumbs

½ cup tomato juice
1 teaspoon salt
⅛ teaspoon pepper
½ teaspoon grated nutmeg
½ teaspoon powdered allspice
½ teaspoon powdered cinnamon

Combine all ingredients and lightly shape into a loaf. Bake in 350° oven about 75 minutes. Serves eight.

HOT DRIED BEEF

¼ pound dried beef, ground
1 clove garlic, minced
1 cup canned tomato sauce

½ cup soy curd (page 35)
1 egg, slightly beaten
Pepper

Combine all ingredients in top of double boiler. Cook until well blended, stirring constantly. Serve on toast, seasoned with pepper. Serves four.

ORIENTAL BURGERS

1 tablespoon fresh ginger root,
   finely chopped
2 cloves garlic, finely
   chopped
½ cup onion, finely chopped
2 tablespoons sugar

½ cup soy sauce
¼ cup water
¾ pound ground beef
1 cup soy pulp (page 29)
1 teaspoon salt
6 hamburger buns

In a saucepan, combine ginger root, garlic, onion, sugar, soy sauce, and water. Heat to dissolve sugar. Combine ground beef, soy pulp, and salt. Pour sauce over meat and let stand one or two hours in refrigerator. Spread mixture on bottom half of hamburger buns and broil three minutes. Makes six sandwiches.

CHOW MEIN WITH MEAT

2 cups onions, sliced
4 tablespoons cooking or salad
   oil
2 cups soy sprouts (page 38)
2 cups cooked leftover chicken
   or pork, diced
½ cup canned or fresh mushrooms, chopped

Water
2 tablespoons soy sauce
Salt
2 tablespoons cornstarch
2 tablespoons water
Chow mein noodles

Stir-fry onion in oil until transparent and golden. Add soy sprouts, meat, and mushrooms (with liquid, if canned). Add cooking liquid from soy sprouts and enough water to barely cover. Cover and simmer 10 minutes. Season to taste with soy sauce and salt. Blend cornstarch and water and stir in. Cook over low heat, stirring constantly, until thickened slightly. Serve over chow mein noodles. Serves six.

Soybean Vegetable Dishes

Vegetables don't *have* to be tiresome. They just seem that way in the hands of some cooks who give the poor things short shrift. A pinch of salt and a dollop of butter just don't do the trick, unless we're talking about artichokes. Throw them together in different combinations, casserole style, fortify them with our friend the soybean, and, voilà! Vegetables of another stripe!

BARLEY-SOYBEAN CASSEROLE

½ cup whole barley, hulled
  and soaked
2 cups cooked soybeans (page 24)
2 medium potatoes, cooked
  and diced
2 ribs celery, diced
2 medium onions, sliced
1 carrot, grated
1 cup cooked or canned
  tomatoes

1 teaspoon salt
3 tablespoons nutritional yeast
½ teaspoon dried savory,
  crushed
½ teaspoon dried chervil,
  crushed
1 teaspoon dill seeds, crushed
1 bay leaf

Blend all ingredients. Bake in oiled casserole 30 minutes in 350° oven. Serves six.

STUFFED ZUCCHINI

2 large zucchini
1 onion, finely chopped
2 ribs celery, chopped
2 tablespoons cooking or salad
   oil

2 tablespoons parsley, chopped
⅛ teaspoon savory
½ teaspoon salt
½ cup cooked soy grits (page 28)
2 tablespoons water

Scoop pulp from zucchini, leaving shell ¼-inch thick. Chop pulp and combine with onion and celery. Sauté in oil until onion is transparent. Add parsley, savory, salt, soy grits, and water. Pile into zucchini shells. Bake in 350° oven 15 minutes, until lightly browned. Serves six.

EGGPLANT CASSEROLE

1 tablespoon onion, chopped
½ cup green pepper, diced
2 tablespoons cooking or salad
   oil
1 medium eggplant, peeled and
   diced
3 tablespoons wheat flour
1 teaspoon vegetable broth
   granules or powder
1 cup water

2 tablespoons parsley, chopped
¼ teaspoon dried sage,
   crumbled
1 cup water
½ pound cooked soy curd
   (page 35), sliced
2 medium tomatoes, sliced
Salt and pepper
¼ cup grated Parmesan cheese

Sauté onion and green pepper in oil over low heat until lightly browned. Add eggplant and stir-fry two minutes. Stir in flour, broth granules, parsley, and sage, then gradually add water and cook, stirring constantly, until smooth and thickened. Pour one-half the mixture into a greased casserole. Add a layer of soy curd, then a layer of tomato slices. Sprinkle each layer with salt and pepper. Repeat. Top with Parmesan cheese. Bake in 350° oven 30 to 40 minutes, until tomatoes are lightly browned. Serves six.

SOY-EGGPLANT BAKE

2 tablespoons cooking
  or salad oil
2 cups soy grits (page 27)
2 cloves garlic, minced
¼ cup tomato paste
2 cups cooked or canned
  tomatoes
1 cup water
1 teaspoon salt

⅛ teaspooon pepper
1 medium eggplant
2 eggs, slightly beaten
½ cup roasted soy grits (page
  48), finely ground
½ pound Mozzarella cheese,
  thinly sliced
½ cup grated Parmesan cheese
½ teaspoon powdered oregano

Heat oil in heavy skillet. Add uncooked soy grits and garlic and brown lightly, stirring constantly. Add tomato paste, tomatoes, water, salt, and pepper. Cook over low heat 30 minutes, until sauce is thickened. Cut eggplant in ½-inch slices and peel slices. Sprinkle each slice with salt and dip in beaten egg, then in ground, roasted soy grits. In a large shallow baking dish, arrange layers of eggplant, Mozzarella cheese, and tomato sauce. Top with grated Parmesan, then oregano. Bake 30 minutes in 350° oven. Makes six servings.

BEAN-CURD CASSEROLE

2 cups cooked soybeans (page 24)
2 cups cooked or canned corn
1 cup pressed soy curd (page 37),
  chopped
2 cups canned or cooked
  tomatoes
½ cup celery, sliced

1 small onion, chopped
2 teaspoons vegetable broth
  granules or powder
1 teaspoon salt
2 tablespoons butter or
  margarine, melted
½ cup bread or cracker crumbs

Arrange alternate layers of soybeans, corn, and soy curd in casserole. Drain tomatoes, reserving juice. Add drained tomatoes to casserole and mix juice with celery, onion, vegetable broth powder, and salt. Pour over all. Top with buttered crumbs and bake 30 minutes in 350° oven. Serves six to eight.

144

SPROUTS AND TOMATOES

3 cups soy sprouts (page 38)
1½ cups water
½ teaspoon salt
½ cup celery, chopped
¼ cup onion, chopped
¼ cup green pepper, chopped

2 tablespoons cooking or salad oil
2 cups cooked or canned tomatoes
1 bay leaf
1 teaspoon sugar
½ teaspoon salt

Cook soy sprouts five minutes in salted water. Drain and set aside. In another saucepan, sauté celery, onion and green pepper in oil until onion is transparent. Add tomatoes, bay leaf, sugar, salt, and sprouts. Cook 10 minutes. Remove bay leaf. Serves six.

SAUTÉED SPROUTS

2 tablespoons cooking or salad oil
1 clove garlic, chopped

1 small green onion, chopped
2 cups soy sprouts (page 38)
Broth or vegetable juice

Heat oil. Sauté garlic and onion until transparent. Add soy sprouts and stir-fry three minutes. Moisten with a small amount of broth or vegetable juice.

FRESH SOY SPROUTS

5 cups soy sprouts (page 38)
1½ cups boiling water
1 teaspoon salt

1 tablespoon cornstarch
1 tablespoon butter or margarine
1 tablespoon lemon juice

Simmer soy sprouts in covered saucepan in water and salt 10 minutes, or until crisply tender. Drain, reserving liquid. In another saucepan, blend cornstarch with reserved liquid. Add butter or margarine and cook until thickened. Pour over soy sprouts. Add lemon juice and stir well. Serves six.

STEAMED SOY SPROUTS

2 cups soy sprouts (page 38)   ½ teaspoon salt
3 tablespoons water   Dash of paprika

Steam soy sprouts in boiling water over low heat 10 minutes. Season
with salt and paprika. Serves four to six.

SOY-ZUCCHINI PATTIES

4 unpeeled zucchinis, finely   1 teaspoon salt
  chopped   ½ cup cooked soy grits (page 28)
4 green onions, chopped   3 tablespoons cooking or salad
2 eggs   oil

Combine all ingredients except oil. Beat well and drop by tablespoonfuls into hot oil. Fry until golden on both sides. Serves four to six.

SOY CURD WITH
TOMATOES

1 large onion, thinly sliced   ½ pound pressed soy curd (page
1 cup celery, thinly sliced   37), cut in thin strips
2 tablespoons cooking or salad   2 tablespoons wheat flour
  oil   2 tablespoons water
4 cups cooked or canned   ½ teaspoon salt
  tomatoes   ⅛ teaspoon pepper

Sauté onion and celery in oil until onion is transparent. Add tomatoes
and simmer 10 minutes. Add soy curd, salt, and pepper, and simmer
over low heat 20 minutes, being careful not to break up curd in stirring.
Blend flour and water and add, stirring gently until slightly thickened.
Season to taste with salt and pepper. Serves six.

SOYBEAN SUCCOTASH

1 cup fresh soybeans, (page 22)
  cooked and drained
1 cup canned or fresh corn

1 tablespoon canned or fresh
  pimiento, chopped
Salt and pepper to taste.

Combine and simmer 10 minutes. Serves four.

SOY-VEGETABLE LOAF

2 cups cooked dried soybeans
  (page 24), drained
3 carrots
1 small onion
2 ribs celery
2 teaspoons vegetable broth
  granules or powder

¾ cup canned or cooked
  tomatoes, chopped
1 teaspoon salt
1½ cups dry bread crumbs

Chop soybeans, carrots, onion, and celery, using coarse blade of food chopper or running quickly through blender. Add vegetable broth granules, tomatoes, salt, and bread crumbs. Pack into an oiled loaf pan and bake 45 minutes in 350° oven. Serves four to six.

PIMIENTO SOYBEANS

1 tablespoon onion, finely
  chopped
2 tablespoons butter or
  margarine
1 cup water
½ teaspoon salt
1 tablespoon cornstarch

2 teaspoons water
1 tablespoon lemon juice
4 cups cooked fresh soybeans
  (page 22)
½ cup fresh or canned pimiento,
  cut in thin strips

Sauté onion in butter or margarine until transparent. Stir in 1 cup water and salt. Dissolve cornstarch in 2 teaspoons water and add. Cook until thickened. Add lemon juice, soybeans and pimiento. Reheat. Makes six servings.

Supper Dishes

Brunches, lunches, and après-ski parties are growing in popularity for informal entertaining these days. You can win a lot of Brownie points with a platterful of Spinach Crêpes. Even the young will forget the spinach all nestled inside its pancake envelope. This and many other recipes can go it alone or play a duet with a tray of colorful raw vegetables. Each one of the following supper dishes has its own form of soy supplement.

CURRIED VEGETABLE PIE

Soy pie crust (page 80), doubled
2 cups cooked soy grits (page 28)
2 teaspoons vegetable broth
  granules or powder
½ cup onion, chopped
⅓ cup apple, peeled and
  shredded
4 teaspoons curry powder
2 tablespoons cooking or salad
  oil

1 cup mixed vegetable juice
  cocktail
1 cup frozen or canned mixed
  vegetables, cooked and
  drained
¼ cup soy milk yogurt (page 54)
2 tablespoons chutney or sweet
  pickles, chopped
1 tablespoon cornstarch
½ teaspoon salt

Line a 9-inch pie pan with one-half the pie crust dough. Roll out remaining dough for top crust and cover with waxed paper. In a heavy skillet, cook soy grits, vegetable broth granules, onion, apple, and curry powder in oil until onion is tender. Stir in vegetable juice, vegetables, yogurt, chutney, cornstarch, and salt. Pour into pie shell and cover with top crust. Seal and flute edges. Make slits in top to allow steam to escape. Bake in 375° oven about 30 minutes, or until crust is browned. Let set 15 minutes before serving. Serves six.

MOCK PIROSHKI

Filling:

½ teaspoon dried savory
1 tablespoon hot water
1 egg, beaten
1 teaspoon vegetable broth
   granules or powder
¼ teaspoon salt
Pepper
½ teaspoon thyme
¼ teaspoon dried sage,
   crumbled
½ teaspoon onion, grated
½ cup fresh or canned
   mushrooms, chopped

1½ cups bread crumbs
2 tablespoons butter or
   margarine
¾ cup soy nuts (page 47),
   coarsely ground
3 tablespoons butter or
   margarine
3 tablespoons wheat flour
¼ teaspoon salt
1 cup soy milk (page 33)

Dissolve savory in hot water. Add beaten egg, broth granules, salt, pepper, thyme, sage, onion, mushrooms, bread crumbs, butter or margarine, and soy nuts. Set aside. In a saucepan, melt 3 tablespoons butter or margarine. Stir in flour and salt. Gradually add soy milk and cook, stirring constantly, until thick. Stir in first mixture.

Pastry:

1¾ cups wheat flour
¼ cup soy flour (page 29)

⅔ cup solid vegetable shortening
Cold soy milk (page 33)

Combine wheat flour, soy flour and salt. Using a pastry blender or two knives, cut in shortening to the consistency of coarse meal. Add just enough soy milk to make a dough. Roll out and cut into six squares. Cut each square in half.

To Assemble:

Divide filling evenly among the twelve pastry pieces. Fold each pastry over filling and seal edges. Pierce tops with tines of a fork and bake in 400° oven 20 minutes. Makes twelve pastries.

MEATLESS EGG ROLLS

Filling:

1 cup soy sprouts (page 38)
4 green onions, chopped
1 cup celery, chopped
1 cup soy curd (page 35)
½ cup fresh or canned
    mushrooms, sliced

2 tablespoons soy sauce
1 teaspoon salt
2 eggs

Combine all ingredients and fill 24 egg roll skins.

Skins:

1¼ cup wheat flour
¼ cup soy flour (page 29)
1½ cups water

3 eggs
1 teaspoon salt

Combine to make a thin batter. Heat a small skillet and lightly coat with oil. Pour in 2 tablespoons batter. Tilt pan so that batter coats skillet, then pour surplus back into bowl. Cook to set skin, but do not brown. Remove to a towel. Repeat to use all batter.

To Assemble:

Combine 1 tablespoon wheat flour and 2 tablespoons water. Set aside. Place 2 tablespoons filling on each skin. Brush edge all around with flour-water mixture. Fold in two sides, then fold in third side and roll into a tight roll. Seal edges. Fry in deep, hot oil until crisp and browned on all sides. Makes twenty-four egg rolls.

ONION CURD PIE

1½ cups cracker crumbs
½ cup butter or margarine,
   melted
2½ cups onions, sliced thin
2 tablespoons butter or
   margarine

1½ cups soy milk (page 33),
   heated
3 eggs, slightly beaten
1 teaspoon salt
¼ teaspoon pepper
½ pound pressed soy curd
   (page 37), chopped

Combine crumbs and ½ cup melted butter or margarine. Blend thoroughly and press evenly in oiled, deep 9-inch pie pan. Sauté onions in 2 tablespoons butter or margarine until lightly browned. Spread over bottom of crumb crust. Add hot soy milk to eggs, stirring constantly. Add salt, pepper and soy curd. Pour over onions. Bake in 325° oven 40 to 45 minutes, until a silver knife inserted in the center comes out clean. Serves six.

POTATO-CURD BAKE

1 medium onion, chopped
½ medium green pepper,
   chopped
2 tablespoons cooking or salad
   oil
2 cups tomato sauce
2 cups cooked or canned
   tomatoes

1 teaspoon salt
1 teaspoon sugar
1 teaspoon dried basil
1 teaspoon powdered oregano
2 eggs
1 cup soy curd (page 35)
4 medium potatoes, sliced
½ cup Cheddar cheese, grated

Sauté onion and green pepper in oil. Add tomato sauce, tomatoes, salt, sugar, basil, and oregano. Simmer 10 minutes. Blend the eggs and soy curd. Arrange layers in a 9 × 12-inch baking dish, beginning with one third of the tomato sauce and one third of the soy curd. Top with one third of the potato slices, then repeat, ending with the curd-egg mixture. Sprinkle with grated cheese. Cover and bake 45 minutes in 400° oven. Serves eight.

ITALIAN LOAF

⅓ cup cornstarch
⅔ cup whole wheat flour
1 cup soy flour (page 29)
1½ teaspoons salt
¼ teaspoon dried sage,
    crumbled

¼ teaspoon dried marjoram
1 cup tomato paste
4 tablespoons cooking or salad
    oil
1 cup warm water

Combine dry ingredients, including seasonings. Mix tomato paste, oil, and water. Add gradually to dry ingredients, beating to a smooth paste. Pour into clean, oiled soup cans and cook in pressure pan at 15 pounds pressure for 1½ hours. Let set 10 minutes to firm up, then remove from cans by cutting off bottoms of cans and forcing loaf out through the top. Decorate tops with catsup and slice while hot. Serves four to six.

FRESH SOYBEAN PIE

1 cup cooked fresh soybeans
    (page 22)
1 cup cooked carrots, chopped
1 cup cooked or canned green
    beans
2 cups cooked potatoes, chopped
3 tablespoons butter or
    margarine

1 medium onion, minced
3 tablespoons wheat flour
2 cups soy milk (page 33)
½ teaspoon salt
½ teaspoon celery salt
1 cup Cheddar cheese, grated
½ cup dry bread or cracker
    crumbs

Arrange first four ingredients in layers in a casserole. Melt butter or margarine in a saucepan. Add onion and cook until transparent. Stir in flour and brown until golden. Gradually add soy milk and cook, stirring constantly, until thickened. Add salt, celery salt, and cheese. Cook until cheese is melted. Pour over vegetables in casserole. Top with crumbs and bake 45 minutes in 350° oven. Serves eight.

SPINACH CRÊPES

¾ cup wheat flour
2 teaspoons sugar
½ teaspoon salt
¾ cup soy milk (page 33)
3 eggs, slightly beaten
2 cups cooked soy grits
   (page 28)
2 tablespoons cooking or salad
   oil
2 teaspoons beef broth granules
   or powder
1 can condensed cream of celery
   soup, undiluted
½ pound (about 1 cup) soy curd
   (page 35)

½ pound fresh spinach, cooked,
   drained and chopped
¼ cup grated Parmesan cheese
¼ teaspoon salt
2 tablespoons butter or
   margarine
½ cup fresh or canned
   mushrooms, sliced
¼ cup green onions, sliced
⅛ teaspoon grated nutmeg
¼ cup water
½ teaspoon lemon juice
Parsley

To Make Crêpes:

Mix flour, sugar, and ½ teaspoon salt. In another bowl, combine soy milk and eggs. Add, all at once, to dry ingredients and beat until smooth. Grease bottom of heavy, medium-sized skillet or griddle while very hot. Cover bottom of pan with a very thin layer of batter and tilt pan so batter is spread evenly. Crêpe should be paper thin. Brown on both sides. As each crêpe is made, cover and set aside in a warm oven. Makes twelve crêpes.

Filling:

Lightly brown cooked soy grits in 2 tablespoons oil. Add beef broth granules, 2 tablespoons undiluted soup, soy curd, chopped spinach, Parmesan cheese, and ¼ teaspoon salt. Spoon about ¼ cup filling on each warm crêpe and roll up, jelly-roll fashion. Arrange filled crêpes, seam-side down, in shallow oven dish in which it will be served. Cover and keep warm in a low oven.

Sauce:

To make sauce, melt 2 tablespoons butter or margarine in a small saucepan. Add mushrooms, onion slices, and nutmeg. Cook until tender. Add remainder of soup, water, and lemon juice. Heat just to boiling, stirring frequently. Pour hot over crêpes in dish. Garnish with parsley. Serves six.

SOY SPROUTS
AU GRATIN

2 tablespoons cooking or salad oil	Pepper
2 tablespoons wheat flour	3 cups soy sprouts (page 38)
1 cup soy milk (page 33)	¼ cup dry bread crumbs
¾ cup Cheddar cheese, grated	2 tablespoons butter or margarine, melted
Salt	½ teaspoon paprika

Pour oil into medium saucepan. Stir in flour and gradually add soy milk. Stirring constantly, cook until thickened. Add ½ cup cheese and salt and pepper to taste. Stir until cheese is melted. Pour over soy sprouts in an oiled casserole. Blend crumbs, melted butter or margarine, remaining cheese, and paprika. Sprinkle over top of casserole. Bake in 350° oven 15 to 20 minutes. Serves six.

SCALLOPED SOYBEANS

3 cups cooked fresh soybeans (page 22)	½ teaspoon salt
1 small onion, finely chopped	¼ cup boiling water
1 cup celery, diced	½ cup dry bread crumbs
½ green pepper, diced	3 tablespoons butter or margarine, melted
½ cup tomato sauce	

Combine first seven ingredients in 1½-quart casserole. Mix bread crumbs with melted butter or margarine and sprinkle over top. Bake 1½ hours in 350° oven. Serves six.

SANDWICH BAKE

2 tablespoons cooking or salad
  oil
½ pound ground beef
1 cup soy pulp (page 29)
1 tablespoon onion, chopped
2 tablespoons butter or
  margarine
8 slices bread

1 teaspoon prepared mustard
1 cup processed American
  cheese, shredded
3 eggs, beaten
2 cups soy milk (page 33)
¾ teaspoon salt
⅛ teaspoon pepper

Heat oil in heavy skillet. Combine beef, soy pulp, and onion and cook, stirring constantly, until well browned. Meanwhile, butter four slices bread and lightly brown buttered side in a 350° oven. Add mustard to meat mixture and spread over toasted bread slices. Top each with shredded cheese and cover with remaining bread slices. Place in a baking pan in one layer. In a bowl, combine eggs, soy milk, salt, and pepper. Beat well and pour over sandwiches. Bake in 350° oven 45 minutes. Serves four.

SOY CORN FRITTERS

1 cup cooked fresh soybeans
  (page 22), drained
2 cups canned or cooked corn
4 tablespoons wheat flour

1 egg
1 teaspoon salt
1 teaspoon sugar
¼ cup hot oil for frying

Mash cooked soybeans slightly with potato masher. Add corn, flour, egg, salt, and sugar. Blend well. Drop by tablespoonfuls into hot fat and brown on each side. Serves six.

Growing Soybeans for Animal Feeds

Hay or Pasture

Any variety of soybeans may be grown for hay or pasture. In both, the largest amount of nutrients—particularly the protein—is in the leaves, blossoms and pods. The stems supply mostly crude fiber, which is also necessary for animal health.

Soybeans are best planted in rows, even when they are to be used for hay or pasture. Large areas may be planted in furrows, then covered, or planted in drills without plowing, after a corn or wheat crop. Soybeans also may be planted by broadcasting the seed and then harrowing the field to cover the seed.

When soybean seed to be used for hay or forage is broadcast or sown two to three inches apart in rows eight inches apart, you'll need approximately 75 pounds of seed per acre. An acre will provide one to three tons of cured hay.

Hay

When cut and cured at the proper state, soybean hay is an excellent livestock feed. Unlike most hays, it is a good source of protein, a nutrient vital to young and lactating animals. The use of soybean hay reduces the need for protein grain supplements. When cut in full bloom, soybean hay is 19 percent protein, although the quantity is less when cut at this stage.

Soybean hay may be harvested by hand or by machine. For maximum yield, it should be cut when the seeds have begun to set but before the leaves turn yellow. If it is cut too early, before the seeds begin to set, the quantity of hay will be less, since the plant is not yet full size, and the hay will be difficult to cure. If cutting is delayed beyond the ideal stage, the leaves may drop during curing and the stems will be tough.

Choose a day that is dry and sunny, with no rain in the forecast. Wait until the sun has dried the morning dew, then cut the plants just above ground level. Let the plants lie in the row until the leaves are thoroughly wilted, but not brittle, usually the next day. Again wait until the dew has dried off. The plants are then raked into windrows or shocks and allowed to cure four to five days.

When dry, the hay may be baled or it may be stored loose, under cover. In drier climates, it is possible to stack dried hay in the field and cover with a tarp or plastic. At all stages, the hay should be handled carefully to avoid dropping the leaves which are a nutritious part of the plant.

Soybean hay takes about twice as long to cure as other hays. However, it is less susceptible to damage from rain during this period. If the leaves are kept intact, soybean hay can withstand two or three rains during drying without much loss of nutrients. It also may be stored for long periods without loss.

Soybean hay compares to alfalfa hay in nutrients. Because it is slightly laxative, limited amounts should be given, especially for the first few weeks of hay feeding. Feeding may be limited by mixing soybean hay half and half with grass hay or by feeding only as much as can be eaten in a reasonable time. Soybean hay should not be kept before an animal at all times.

Pasture

Soybeans may be planted as for hay and used as a pasture plant. It is especially valuable for young, growing animals because of its high protein content and its ability to grow well even during drought conditions once the crop is established. Soybeans often succeed as a pasture plant where other crops fail.

Because the soil-enriching nodules are in the roots of the plant, live-stock may be pastured in the field when the soybean crop is grown for soil improvement. In this way, the nodules enrich the soil, the livestock benefits from the protein-rich feed, and the manure from the animal fertilizes the field.

Soybeans are especially valuable in this double role of soil builder and livestock feed when used as a rotation crop with grains. They also may be interplanted with grasses and used for hay or with corn and sorghum, then harvested or pastured for a well-balanced, ready-mixed grain feed. Grains interplanted or rotated with soybeans often have a better yield because of the plentiful supply of nitrogen in the soil.

Silage

Soybean plants also may provide needed protein in silage for livestock. A simple method of planting for silage is to mix the seeds, then plant them together—20 pounds to the acre in rows 40 inches apart. The crop is harvested when it is green and succulent, then chopped for silage.

Here are two seed mixtures that work well with this method of planting and harvesting:

3 parts corn	*Or:*	3 parts corn
1 part sorghum		1 part sunflowers
1 part soybeans		1 part soybeans

Grain Supplement

To grow soybeans for use as a protein supplement in livestock grain feeding, follow the growing suggestions on pages 6 through 14, allowing 60 pounds of seed to the acre and planting in rows 18 to 20 inches apart. The amount to plant will depend on many factors—the number and type of livestock and the grain mixture to be used. In general, you can expect a yield of up to 20 bushels to the acre under good conditions and harvesting by hand. This is far more than is needed for a few farm animals.

If soybeans are being planted in a field for the first time or if it has been more than two years since they were grown there, the seed will need to be inoculated as described on pages 12 and 13.

Harvest the dried beans by machine or by hand, threshing them according to the instructions on pages 15 to 16. Soybeans for livestock use may be dried and stored the same as for human food.

Raw soybeans may be substituted for commercial soybean meal in the grain feedings of ruminants. Because of the anti-trypsin factor in soybeans, one-stomach animals cannot completely utilize soybeans, making them an uneconomical feed if the soybeans must be purchased.

However, beef and dairy cattle, goats, and sheep may be fed raw soybeans up to 20 percent of the grain ration. The one-stomach animals— hogs, rabbits, chickens, etc.—may be fed grain mixtures containing up to 10 percent raw soybeans supplemented by soybean hay or another form of protein such as non-fat dry milk, fed as a liquid or in dry form mixed with the grain.

Both types of livestock prefer soybeans cooked, either simmered or roasted. This is not always possible, but it is not too difficult to spread the soybeans on shallow trays and roast them lightly in a 300° oven for a few animals. Roasted soybeans can be stored in a closed container just as are raw soybeans. Simmered soybeans are well liked by livestock but must be refrigerated or used within a few hours. In most cases, it is preferable to feed the soybeans raw. Like humans, the animals will become accustomed to the taste if the soybeans are introduced slowly into the grain mixture.

Whether soybeans are used raw or cooked, the addition of a small amount of molasses, liquid or dry, will increase the palatability of the grain mixture.

In mixing grain for livestock feeding, it is always more economical to use homegrown grains or those available in your locality at lowest cost.

Because of its contribution of Vitamin A and its low cost, corn usually is the basis for most livestock grain mixtures. It also is well liked by most livestock. The other grains—wheat, oats, barley, buckwheat, millet, or rye—may be used interchangably, according to their cost and availability locally. In areas or at times when corn is expensive, Vitamin A may be supplied through the feeding of raw pumpkin or yellow squash, both of which are also well liked by livestock.

Except where specified, the grains should not be ground but should be mixed whole. Most animals—even young animals—are capable of grinding whole grain satisfactorily. Grains which are stored after grinding tend to turn rancid quickly in warm weather and the rancid oils may be toxic to animals. Ground grains also lose flavor and vitamins rapidly.

Where grinding is necessary, it is best to mix the grain in small batches, no more than will be used in a week. Freshly ground grain is not only better for livestock, it is also much more palatable to them.

The following livestock grain mixtures are to be used not as inflexible rules, but as guidelines for creating your own livestock feeds from homegrown soybeans and other grains available in your area.

Suggested Grain Mixtures
Using Whole Soybeans

(For animals on pasture or green hay)

FOR DAIRY ANIMALS IN MILK:
(cows or goats)

75 pounds whole, shelled
    corn
75 pounds oats
25 pounds wheat

25 pounds soybeans
2 pounds salt
10 pounds dry molasses

For a Young Calf:

(1 to 6 weeks old)

30 pounds whole shelled corn
10 pounds oats
10 pounds wheat
10 pounds soybeans

5 pounds bone meal or dry milk
   solids
1 cup salt
4 ounces cod liver oil

For a Young Growing Calf:

(6 weeks to 6 months)

30 pounds whole shelled corn
30 pounds oats
30 pounds wheat

10 pounds soybeans
20 pounds dry milk solids

For Older Calves:

100 pounds of whole, shelled
   corn
50 pounds oats

25 pounds wheat
25 pounds soybeans
2 pounds salt

For Horses and Mules:

(active or working animals)

80 pounds oats
50 pounds corn

10 pounds soybeans
2 pounds molasses

For Brood Mares with Nursing Foals:

30 pounds oats
30 pounds wheat

10 pounds soybeans
2 pounds molasses

For Sheep and Goats not in Milk:

50 pounds whole, shelled
   corn
20 pounds oats

20 pounds wheat
10 pounds soybeans
2 pounds molasses

For Fattening Lambs:

30 pounds whole, shelled
   corn

2 pounds soybeans
1 pound molasses

(This mixture should be coarsely ground for better consumption.)

For Hogs:

65 pounds whole shelled corn
15 pounds wheat

15 pounds soybeans (preferably
   roasted)

(If soybeans are roasted, grains may be mixed whole. If soybeans are used raw, grains should be ground for better consumption.)

For Growing Chickens:

30 pounds ground corn
5 pounds ground soybeans
   (preferably roasted)
10 pounds rolled oats

10 pounds dry milk solids
1 cup salt
½ cup cod liver oil

For Laying Hens:

40 pounds ground corn
7 pounds ground soybeans
10 pounds rolled oats

5 pounds dry milk solids
1 cup salt

For Rabbits:

10 pounds ground corn	10 pounds rolled oats
2 pounds ground soybeans	¼ cup salt
(preferably roasted)	½ cup dry molasses

Homemade Dog and Cat Food

As with the livestock feeds, it is best to use those foods which are cheapest and most easily obtained to provide the best food at the least cost for household pets.

In the following recipe, for instance, carrots are used because they are well liked by the animals, and are easily grown and stored. (Carrots may be covered with bales of straw and kept right in the garden all winter in most climates.) Carrots also provide a good amount of Vitamin A, a nutrient needed by dogs and cats. Other vegetables, especially yellow vegetables such as pumpkin or squash, may be substituted.

Being carnivorous animals, dogs and cats need a larger amount of protein than do the grazing animals, and some animal meat is advisable. Meat or animal by-products often are available very cheaply or free of charge at butchering plants or on the farm at butchering time. Even the average kitchen often has leftover scraps of meat, fat, and gravy which might be frozen and saved for dog or cat food-making time.

If you prefer not to use meat in your pet food, commercial beef or chicken-flavored broth powder or granules may be used for flavoring.

In this case you'll need to add cooking or salad oil to replace the animal fat called for.

The following recipe makes enough dog or cat food to feed a medium-size dog for about one week. For smaller dogs or for cats, you may wish to store part of it in the freezer. For larger animals, the recipe may be doubled or tripled.

DOG OR CAT FOOD

4 cups cooked, mashed soybeans
2 cups cooked, mashed carrots
1 cup non-fat dry milk powder
½ cup leftover animal fat

1 cup cooked meat or animal by-products (beef heart, liver, fish, tongue, kidney, even pressure-cooked chicken bones which are then run through the blender.)
Broth from cooked meat

Combine first five ingredients. Add enough broth to moisten well.

As a moist pet food, this may be fed as is. If larger batches are made, such as at butchering time, it may be packaged and stored in the freezer or canned by processing pints 65 minutes at 10 pounds pressure in a pressure canner.

To make a dry pet food which is more easily stored and used, add enough liquid to make a batter-type mixture to recipe above. Pour it into a well-greased shallow pan or pans and bake one hour in a 325° oven, or until firm and lightly browned. Allow to cool in the pan, then cut it into small pieces with a knife, crumble with the hands or run through a grinder. Once again spread out on cookie sheets or the shallow pans and toast in a 250° oven two hours, or until thoroughly dried. Feed as is or with water or milk. May be stored in plastic bags on the pantry shelf or in the freezer.

Using Soybeans for Soil Improvement

Just as soybeans provide protein for human and animal foods, so do they provide soil protein—nitrogen.

The little nodules on the root of the soybean plants are air pockets in which the nitrogen-fixing bacteria go about their work of converting nitrogen from the air into plant nitrogen.

So efficient are these little workers that they always manufacture more nitrogen than the soybean plant can use. The surplus—soil nourishment to rival that of commercial fertilizers—is left in the soil for the next crop.

It's possible to take advantage of this nitrogen-fixing ability of the soybean and to use the plant not only for human food and animal feed but also to nourish the soil in which it grows.

The process is called *green manuring* because the green plants are fed to the soil to provide nourishment just as manure does when used as a fertilizer.

The principle is simple. As soon as the garden crop is harvested in the summer, soybean seed is planted in that area. For this purpose, the late-bearing, bush varieties are best.

All through the fall rains and the cool weather, the plants are allowed to grow. There's no weeding or worrying about them.

Then, when the plant reaches the height of its green, lush maturity, before its pods form, it is tilled under and its nutrients fertilize the soil. At this stage, the tender plants rot quickly. Within six weeks—long before spring—they will have disappeared into the soil.

In some climates it may be desirable to wait until spring to plow under the plants, but it will be necessary to wait six weeks while the plants rot before planting the spring garden.

The soybean has several advantages over other plants which are used as a green manure crop. Its root nodules produce nitrogen, and it produces a great deal more leaves and pods and stalks than many crops— about twice as much as cowpeas or clover or alfalfa. Unlike clover, which can become something of a nuisance in the garden, soybeans will not come back in the spring from their roots.

The benefits of green manuring to the soil are phenominal. An acre of soybeans can provide approximately 175 pounds of nitrogen, 115 pounds of potash and 45 pounds of phosphoric acid. All for 60 to 75 pounds of seed which you can grow yourself.

Even if your land or your circumstances won't allow you to use soybeans as a green manure crop, you can take advantage of the nitrogen-rich roots of the plants in the garden by tilling the roots under as soon as the beans are harvested. The soybean straw left on top of the garden when the beans are taken, plus any pods you return to the garden, will provide an appreciable amount of nitrogen, potash and phosphoric acid.

Soybeans in Landscaping

Soybeans also can be used to nurture and shade plantings around the house while lending their own bushy green landscaping touches.

For instance, you can plant a circle of soybeans around newly-planted trees and shrubs to shade their root area and keep the soil from drying out, to feed the area with nitrogen and as a warning barrier to protect the plant from trampling feet or lawnmowers.

You can plant soybeans on a steep bank as a nurse crop for grass seed. Plant a "fence" of soybeans around an exposed flower bed until the flowers are established, as a marker between vegetable and flower gardens, along the sidewalk to discourage trespassing on newly planted grass, as a quick cover for hard-to-plant areas, to keep bare soil from washing away.

The nice thing about using soybeans in landscaping is that they won't spread. Not only are they a pretty plant all through their growing and blossoming stages, but when their task is through, you can just pluck off the soybean pods, cut off the plants, and add them to the garden or to the compost heap. The accommodating soybean not only becomes almost anything you want it to, it also grows only where you want it to.

Index

Other Garden Way Books
You Will Enjoy

The Complete Guide to Growing Berries & Grapes, by Louise Riotte. 142 pages, quality paperback, $4.95. What to plant where, when, and exactly how.

Profitable Herb Growing at Home, by Betty E.M. Jacobs. 240 pages, quality paperback, $5.95. The perfect book for those who wish to expand a home herb garden into a money-making country sideline.

What Every Gardener Should Know About Earthworms, by Dr. Henry Hopp. 40 pages, quality paperback, $1.50. The benefits of earthworms in making richer soils and bigger crops.

Secrets of Companion Planting for Successful Gardening, by Louise Riotte. 226 pages, quality paperback, $5.95, hardcover, $8.95. For bigger, more luscious crops.

Down-to-Earth Vegetable Gardening Know-How, featuring Dick Raymond. 160 pages, deluxe illustrated paperback, $5.95. Special durable cover edition, $7.95. A treasury of complete vegetable gardening information.

Keeping the Harvest: Home Storage of Vegetables & Fruits, by Nancy Thurber and Gretchen Mead. 208 pages, deluxe illustrated paperback, $5.95; spiral edition, $6.95; hardcover, $9.95. The very best of the food storage books.

Vegetable Gardening Handbook, by Roger Griffith. 120 pages, spiral bound, $3.95. Take it into your garden, for information and for your own record book.

172

THE SOYBEAN BOOK

Dwarf Fruit Trees for the Home Gardener, by Lawrence Southwick. 118 pages, quality paperback, $3.95; hardcover, $5.95. All you need to know to start a home orchard on a small plot.

Improving Garden Soil With Green Manures, by Dick Raymond and Richard Alther. 48 pages, paperback edition, $2.50. An illustrated, no-frills handbook that shows the tremendous difference green manures can make in improving your garden.

Let It Rot!, by Stu Campbell. 152 pages, quality paperback, $3.95. Homemade fertilizers for a healthier garden.

Cash from Your Garden, by David Lynch. 208 pages, 5½ × 8½, illustrated, quality paperback $3.95. The book that tells how to turn your garden produce into extra income.

The Home Gardener's Cookbook, by Marjorie Blanchard. 192 pages, quality paperback, $4.95; hardcover, $6.95. Mouth-watering recipes using your garden produce.

Treasured Recipes from Early New England Kitchens, by Marjorie Blanchard. 144 pages, quality paperback, $4.95; hardcover $8.95. Yesterday's favorite recipes, adapted to today's kitchens.

Woodstove Cookery: At Home on the Range, by Jane Cooper. 204 pages, illustrated by Sherry Streeter, $5.95. A warm and friendly introduction to buying, installing, and cooking on a kitchen wood-burning range, with hundreds of time-tested recipes.

Homemade: 101 Easy-to-Make Things for Your Garden, Home or Farm, by Ken Braren and Roger Griffith. 176 pages, deluxe illustrated paperback, $6.95. A wonderful collection of simple projects for the home carpenter.

These books are available at your bookstore, or may be ordered directly from Garden Way Publishing, Department 171X, Charlotte, VT 05445. If order is less than $10, please add 60¢ postage and handling.